SpringerBriefs in Business

For further volumes:
http://www.springer.com/series/8860

Caroline M. Sipp • Elias G. Carayannis

Real Options and Strategic Technology Venturing

A New Paradigm in Decision Making

Springer

Caroline M. Sipp
Inter-American Development Bank
Washington, DC, USA

Elias G. Carayannis
The George Washington University
Washington, DC, USA

ISSN 2191-5482 ISSN 2191-5490 (electronic)
ISBN 978-1-4614-5813-5 ISBN 978-1-4614-5814-2 (eBook)
DOI 10.1007/978-1-4614-5814-2
Springer New York Heidelberg Dordrecht London

Library of Congress Control Number: 2012951426

Printed on acid-free paper

Springer is part of Springer Science+Business Media (www.springer.com)

Abstract and Keywords

Abstract This book is motivated by the role of real options as a risk management and an uncertainty-filtering methodology that helps minimize downside risk and maximize upside potential of a firm's investments. It seeks to answer "*why, when, and how are real options used in strategic technology venturing*?" This work tests for the role of real options in decision making involving three types of firms in decreasing order of technology dependence – technology-driven (TD) (where the profit is fully dependent on new technology creation and leveraging), technology-based (TB) (where the profit is enabled and supported by technology), and technology-neutral (TN) (where the profit is almost independent of technology). It also deals with strategic and nonstrategic types of decisions driven by real options.

This analysis shows that an environment presenting co-opetitive (simultaneous competition and collaboration) conditions triggers the use of real options (*why*) that serve to transform the position, posture, and propensity of businesses to innovate and thus they coevolve (*when*) into more effective and efficient forms of businesses (co-specialization) (*how*). The simulation presents evidence that embracing risk and uncertainty can increase level of new venture formation and probability of new venture formation. However, it also shows that it should be adapted to the risk profile of the firm and that timing is also a factor to be considered. While increasing risk at firm creation increases both the level and probability of new venture formation, increasing the rate of early adoption does not significantly affect it. For more risk-averse firms, lowering the rate of early adoption actually significantly increases the level of new venture formation as well as the probability of new venture formation. Finally, evidence seems to point that there may be an optimal level of risk inclination, as the results from the two most risk-taking subpopulations are not statistically significant.

A real-options approach allows for a better mitigation and management of risk and therefore increases the return per unit of asset invested. It brings about a higher potential return per unit of asset while minimizing over-, under-, and mis-investing. The systematic use of real options is a possible powerful methodology and device for optimal risk management and risk filter under variable configuration and operationalization of resource allocation. It optimizes the allocation of scarce resources with timing, selection and sequencing decisions, improving cost/benefit and cost/

efficiency (public sector view), and maximization of performance and strategic benefit (private sector view). Although engaging the concepts of real options, this analysis does not focus on a specific investment valuation methodology, but highlights the relationship between knowledge and risk and rather addresses the management of mindsets, as moving toward a systematic conceptualization of real options represents a different paradigm in decision making.

Keywords: knowledge management, innovation, real options, research and development (R&D), risk management, strategic decision making, strategic investments, technology management, technology venturing

Contents

List of Figures

List of Tables

Chapter 1
Conceptual Framework

In today's hypercompetitive and glocalized marketplace, firms need to systematically leverage and arbitrage strategic knowledge to create sustainable entrepreneurship and robust competitiveness (E. G. Carayannis 2008, 2009). The nature and dynamics of choice evolve with time, driven both by intrinsic and extrinsic factors. In this context, co-opetition, coevolution, and co-specialization events and processes, as well as strategic knowledge arbitrage and serendipity (SKARSE™)[1] (E. G. Carayannis 2008) "conspire" to continually recast and reshape the landscape within which choices are made.

Schumpeter, Simon, March, Penrose, Williamson, Nelson, Winter, and others have woven insights into the fabric of our understanding about when, how, and why socio-technical change precipitated by choices made results in successful outputs, outcomes, and impacts. The Schumpeterian cycle of creative destruction seen through the prism of the theory of the growth of the firm, transaction costs economics, innovation diffusion, and real options may help us establish a more clairvoyant perspective, confronting and pushing the risk/uncertainty frontier.

Under traditional investment decision methods (NPV, ROI), risk and uncertainty are treated as negative factors: the higher the risk and uncertainty, the higher the discount rate and the lower the current value of the investment. However, some investments are naturally linked to higher risk and uncertainty depending on the type of investment, industry, or stage of company life considered (e.g., technology investments, biotech, firm inception), yet they could present a good opportunity for investment. What are traditionally clear investment thresholds become investment decision gray zone that should be looked at in light of real-options thinking. Investment under risk and uncertainty can lead to inappropriate or insufficient valuation and management of risk, resulting in under- or mis-investment, which could be mitigated by deploying real-options methodologies to take more risk more successfully. This is particularly relevant in the aftermath of the latest financial crisis that has strained the budgets of both public and private organizations.

---

[1] SKARSE™: SN 85174407, filed November 11, 2010 by Elias G. Carayannis.

C.M. Sipp and E.G. Carayannis, *Real Options and Strategic Technology Venturing: A New Paradigm in Decision Making*, SpringerBriefs in Business 31,
DOI 10.1007/978-1-4614-5814-2_1,

1 Motivation

This book is motivated by the role of real options as a risk management and an uncertainty-filtering methodology that helps minimize downside risk and maximize upside potential of a firm's investments. Firms evolve in an uncertain and dynamic environment in which they use the "new knowledge [they] derived through the healthy balance between competition and cooperation involving employees and business partners" (*strategic knowledge co-opetition*, E. G. Carayannis 2009) in the definition of their real options. These real options serve as the basis for their decision making so as to reap the full benefits of the flexibility embedded in their investments. By the exercise of their options, firms have changed the parameters of their previously temporarily stable ecosystem, resulting in a now unstable environment. Having completed the co-opetition process, firms create "new knowledge through a series of interactions and changes at various levels of the organization, spurred by the co-generation and complementary nature of that knowledge," what Carayannis (ibid) coined *strategic knowledge co-evolution*. Through innovation, they also undergo *strategic knowledge co-specialization,* "learning and knowledge which encourages individuals or groups to expand their roles into new areas and new domains, in a complementary and mutually-reinforcing fashion" (ibid). Strategic knowledge co-specialization enables firms to develop *sustainable entrepreneurship* (E. G. Carayannis 2008), that is, "the creation of viable, profitable and scalable firms that engender the formation of self-replicating and mutually enhancing innovation networks and knowledge clusters leading towards what we call robust competitiveness." As such, firms are enabled to develop a temporarily "unfair" competitive advantage as they develop knowledge in new areas and exploit them. In this context, firms now evolve in a temporarily stable environment, *sustainable entrepreneurship* (ibid), a "state of economic being and becoming that avails systematic and defensible 'unfair advantages' to the entities that are part of the economy and is built on mutually complementary and reinforcing low-, medium and high technology, public and private sector entities (government agencies, private firms, universities, and non-governmental organizations)."

2 Framework of Analysis

This book seeks to answer the following question, which is motivated by the role of real options as a risk management and uncertainty-filtering methodology that helps minimize downside risk and maximize upside potential:

- Why, when, and how are real options used in strategic technology venturing?

This work tests the role of real options in decision making involving three types of firms in decreasing order of technology-dependence – technology-driven (TD) (where the profit is fully dependent on new technology creation and leveraging), technology-based (TB) (where the profit is enabled and supported by technology), and technology-neutral (TN) (where the profit is almost independent of technology).

It also deals with strategic and nonstrategic types of decisions driven by real options. Finally, this research assumed an environment (business ecosystem), where co-opetitive conditions trigger the use of real options (why), that serve to transform the position, posture, and propensity of businesses to innovate, and thus they coevolve (when) into more effective and efficient forms of businesses (co-specialization) (how).

As such, the book seeks to answer the abovementioned central question by exploring the following aspects:

- Does the use of real options in decision making for technology investments increase in a co-opetitive environment? (why)
- Does the use of real options in decision making for operational, tactical, and strategic technology investments increase with risk and uncertainty? (when)
- Does the use of real options in decision making increase when the technology investment under consideration leads to new business areas (product, services, and/or processes)? (how)

3 Co-opetition, Coevolution, and Co-specialization (C3) and Creative Destruction

Schumpeter (1934) argued that development comes from within an economic system and is related to changes in how the production is organized. Any new combination introduced causes development. This has been validated at the firm level but also at a macro-level by Cimoli and Katz (2003). When analyzing the Latin American economies' shift from state-led development strategies to a market-driven paradigm, Cimoli and Katz concluded that the Schumpeterian creative destruction process played a major role in the sustainability of macro-stability. However, they also conclude that one crucial element of the adjustment process was the continuous entry and exit of firms, showing that co-opetition, coevolution, and co-specialization events as described below cannot only lead to further development but also provide the necessary flexibility when shifting from one state of economic being to another state of economic becoming.

Furthermore, Carayannis (2008) argued that co-opetition, coevolution, and co-specialization events and processes along with diversity (heterogeneity) lead to sustainable entrepreneurship and robust competitiveness, therefore generating changes or disturbances in the ecosystem leading to further development. The view that heterogeneity leads to further development is also shared by Cantner and Hanusch (2001) who when trying to empirically define and measure heterogeneity identified it as follows:

> Heterogeneity is a concept which refers to *the degree of difference within a population of observations*, let that be households, firms, sectors or even regions or countries which differ with respect to their efforts, behaviors and/or success due to – among others – the artefacts they consume or produce, the modes of production they employ, the direction and intensity of innovative activities they pursue, or the organizational setting they choose. This heterogeneity of agents is, on the one hand, considered the *result* of technological change, i.e. of different innovative/imitative/adaptive activities and differential innovative/

> imitative/adaptive success; on the other hand, it also serves as a *source* for further progress in the sense that this heterogeneity puts pressure on technologically backward actors to improve performance when the gaps become too large and on leaders when the gaps become too close; and that it provides for different kinds of learning processes (imitative and adaptive learning, cross-fertilization, etc.). (emphasis in original)

Moreover, in an attempt to laying the groundwork of a theory that would link the role of the firm and economic development, Moran and Ghoshal (1999) provide arguments supporting the illustration of co-opetition, coevolution, and co-specialization processes and diversity below. Moran and Ghoshal base their arguments on Schumpeter's theory that development is created by any new combination of resources but argue that the combination of resources is not enough and that research should focus also on access to new combination of resources and that three conditions should be met: (1) "the deployment must be *enabled* – that is, someone must have access to all the requisite resources to execute the deployment," (2) "the deployment must be *motivated* – that is, someone must benefit from its execution," and (3) "some service must be *perceived* to flow from the deployment – that is, someone must see the deployment as a viable act from which some service is anticipated." Since it is unlikely that all these conditions reside within the same individual, the greater the number of individuals involved, the greater the chance for a new combination to be deployed. As companies are finite elements and vary in size, the probability is high that these conditions may be satisfied across firms, at which point, firms will enter in exchange and not just production of new combinations of resources. This notion of exchange provides strong support for the processes of co-opetition, coevolution, and co-specialization across firms, as illustrated in Fig. 1.1.

3.1 Transitioning from an Ecosystem to the Next Through Co-opetition, Coevolution, and Co-specialization (C3)

Co-opetition, coevolution, and co-specialization and diversity can lead to new ventures, mergers, and acquisitions. Figure 1.1 is an illustration of how co-opetition, coevolution, and co-specialization (C3) events and processes can lead from one ecosystem to the next through mergers.

As shown in Fig. 1.1., both new (N) and existing companies (O) compete in a given ecosystem (ecosystem 1) while going through the organizational cognition spiral (OCS) (E. G. Carayannis 1998) that impacts both the behavior of the company and that of its competitors. Each learning order is a cycle of ignorance of ignorance (I/I), awareness of ignorance (A/I), awareness of awareness (A/A), and ignorance of awareness (I/A) (ibid). As companies go through the first order of learning (learn: develop operational knowledge), then the second order of learning (learning how to learn: develop tactical knowledge), and finally the third-order learning (learning to learn how to learn: develop strategic knowledge) (E. Carayannis 1994, 1998, 2008), they derive new knowledge from competition and cooperation

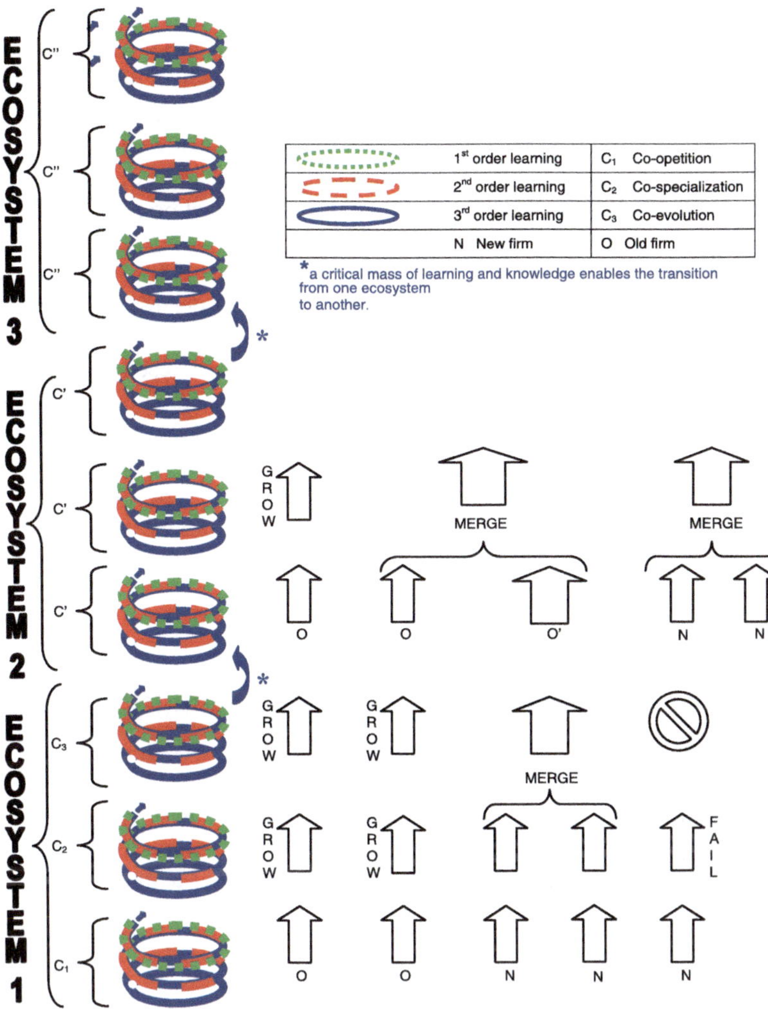

Fig. 1.1 From one ecosystem to the next: organizational cognition spiral (OCS) in C3 (Carayannis and Sipp 2010, p. 74)

and reach a co-opetition stage (C1). As they grow (G), merge (M) or fail (F), and experience the OCS again, they create new knowledge based on complementary knowledge derived from interactions at various level of the organization and reach a stage of coevolution (C2). As they go through a third OCS, they allow groups and individuals to develop knowledge in new areas, and they grow in to a co-specialization stage (C3) (E. G. Carayannis 2008).

At this stage, companies reach a state of robust competitiveness. All the companies are robust or competitive enough to survive and prosper in the ecosystem, the weaker ones having improved or disappeared. In the state of robust competitiveness

(E. G. Carayannis 2009), there is an optimal number of companies leading a relatively stable ecosystem (for some period of time) in the direction they want. This is, however, not an oligopoly. Robust competitiveness is open to innovation and new entrants, leading to sustainable entrepreneurship (ibid).

These "happy accidents" and spontaneous events (E. G. Carayannis 2008), as well as the natural development and change, both absolute and relative, of existing ventures and companies (including the end of a technology or the decline of a market) lead to a critical mass of instability. While in a co-specialization state, the companies have roughly equivalent competencies; the ecosystem is fairly stable. Now that changes have been introduced, there is enough heterogeneity to destroy this "balance of power." Enough heterogeneity generates a need for change resulting in a new balance of power (E. G. Carayannis et al. 2008b). Heterogeneity leads back to co-opetition. It is worth noting, though, that these processes have memory. This series of processes will start over but with a richer knowledge base. Following Schumpeter's principle of creative destruction (1911), both new (N') and existing (O') are back to a stage of co-opetition (C'1), but in a new ecosystem (ecosystem 2).

Co-opetition, coevolution, and co-specialization are dynamic processes describing both phenomena of stasis (more or less stable ecosystems) and change (resulting in the creation of new ecosystems) (E. G. Carayannis 2008).

4 Strategic Knowledge Arbitrage and Serendipity (SKARSE™) as Drivers of Real Options

As Hull (2008) points out, capital investments in real assets often embed options, but these options are difficult to value using traditional capital investment valuation techniques. The real-options approach attempts to deal with this using option-pricing theory as a basis for the analysis.

4.1 Real-Options Analysis Versus "Traditional" Valuation Methods

The "traditional" decision-making techniques are based on the discounted cash flow method (DCF). "The central paradigm for making decisions about large investments is net present value. Unfortunately, it is badly flawed and systematically undervalues every investment opportunities. Why? Because it is based on expected future cash flows and, thereby failing to account for value flexibility" (Copeland and Antikarov 2001).

Discount cash flow methods have three major drawbacks compared to real options: (1) they systematically treat risk and uncertainty as a negative factor (higher discount rate), which managers find is not always the case; (2) they account for predetermined cash flows (do not account for managerial flexibility); and (3) they propose a very limited set of investment timing (invest now or not). Real options embody today's options for tomorrow's growth and account for the value of managerial flexibility.

Luehrman's (1998) description of a portfolio of options as a tomato garden is particularly appealing for its simplicity. A portfolio of options is akin to a tomato garden in that one needs to cultivate it (dynamic stable equilibrium). When one visits his tomato garden, he will invariably encounter tomatoes that are just ripe and good to be harvested (options to exercise) but also some tomatoes that are still too green and need more ripening (options on which to keep an eye on) and tomatoes that will have rotten since his last visit (now dead options). The beautiful and ripe tomatoes will be harvested, but he will come to take care of the still-green tomatoes and come back periodically to check whether they will then be ready for harvest.

Keeping this dynamic portfolio management scheme in mind, real options are about balancing financial imperatives and learning insights and strategy. This strategic balance is a dynamic unstable equilibrium (like walking, "continuum of discontinuities" or imbalances) (E. G. Carayannis 1994, 2000), leading to the creation of a new garden because of changes in the environment but also new varieties of tomatoes planted in the garden.

4.2 *Real Options and Schumpeter's Creative Destruction*

In *Capitalism, Socialism, and Democracy* (1942), Schumpeter posited that capitalism is a continuous evolutionary process, in which the entrepreneur plays a key role since he introduces technological change and innovation, thereby disrupting a state of development in (quasi) equilibrium (Mark I). Creative destruction describes these innovative entries as by so doing, entrepreneurs affect the fundamentals of established businesses. Later, Schumpeter's research postulated that big companies are the main drivers of innovation and technological change (Mark II), as they have a considerable amount of money to invest in research and development. Today, as the public and private sectors and academia work more closely together and as the role of innovation networks' and knowledge clusters in creativity, invention, and innovation has been acknowledged (E. G. Carayannis et al. 2008a), it is more likely to experience a combination of Mark I and Mark II. Furthermore, with the development of a mode 3 system of knowledge production (E. G. Carayannis 2008) that consists of innovation networks and knowledge clusters and that is multilayered, multimodal, multi-nodal, and multilateral, there may even be a shift toward Mark III. Mark III consists of a mode 3 knowledge production system and a Quadruple Innovation Helix (Carayannis and Campbell 2006, 2009). These are cornerstones of a fractal innovation ecosystem architecture (ibid).

Schumpeter further argued that capitalism is by its nature a form or method of economic change and that the results of these changes are known as economic cycles. The use of real-options analysis, however, may not necessarily contemplate Kondratieff cycles as real-options horizon does not typically go beyond 15 years (Copeland and Antikarov 2001). Nonetheless, the same logic may apply to shorter cycles, as the end of an S-curve (see Fig. 5, Carayannis 2009, p. 242). While firms make decisions in a "current state" at time of the decision, they inevitably have to react to changes in their environment (changes in policy, technology, competition),

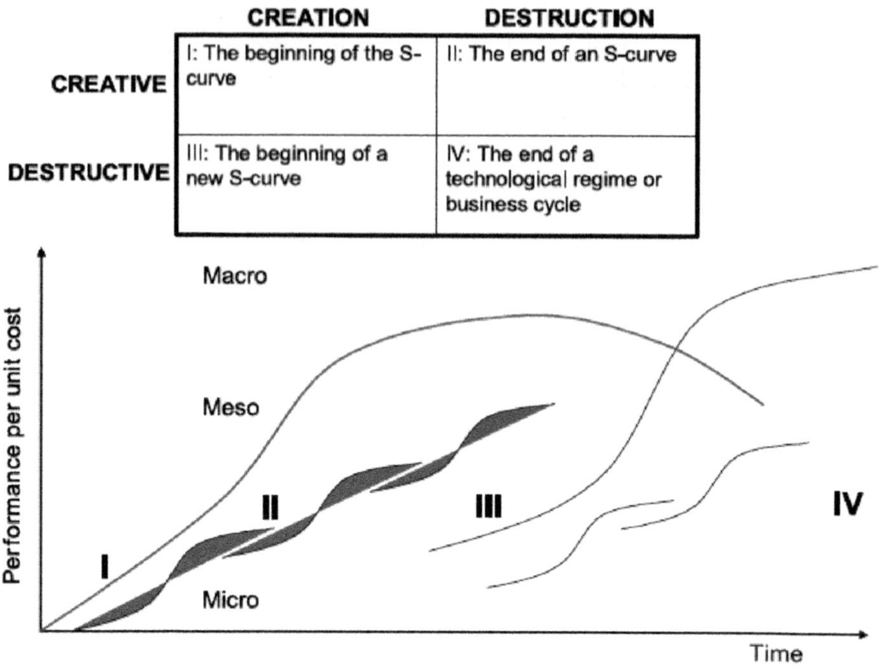

Fig. 1.2 The life cycle of creative destruction (Carayannis 2008, p. 343)

which can either create new opportunities or close some. These options and the corresponding managerial flexibility should be accounted for at the time of decision making, using a valuation method that allows for such variables.

In addition to land, labor, and capital, Schumpeter identified technology and entrepreneurship as inputs of economic activity. Furthermore, Drucker (1968) identified knowledge as the sixth and maybe the most important input. As such, knowledge can be used, leveraged, and recombined as any other production resources and generating development "from within" (Schumpeter). In addition, Carayannis (2008) defined strategic knowledge arbitrage and serendipity as the following:

> [Strategic knowledge arbitrage refers to] the ability to distribute and use specific knowledge for applications other than the intended topic area. More specifically, it refers to the capacity to create, identify, reallocate and recombine knowledge assets more effectively and efficiently to derive, develop and capture non-appropriable, defensible, sustainable and scalable pecuniary benefits. (p. 346)
>
> [Strategic knowledge serendipity refers to] the unintended benefits of enabling knowledge to "spill over" between employees, groups and functional domains ("happy accidents" in learning). More specifically, it describes the capacity to identify, recognize, access and integrate knowledge assets more effectively and efficiently to derive, develop and capture non-appropriable, defensible, sustainable and scalable pecuniary benefits. (p. 346)

Strategic knowledge arbitrage and serendipity forms elements of a knowledge-based real option. Strategic knowledge serendipity defines the terms of the option, and strategic knowledge arbitrage provides managers with the necessary information to decide if, when and how to execute that option (E. G. Carayannis 2008). Consequently, executing knowledge-based real options could be seen as creating a new disequilibrium, leading to creative destruction but also destructive creation, creative creation, and destructive destruction (see Fig. 1.2) (E. G. Carayannis et al. 2008b).

5 The Cognition Spiral and Strategic Themes Viewed Through the Option Lens

Bowman and Hurry (1993) describe four strategic theme based in their real-options-based approach to strategy: sensemaking, resource allocation, strategic positioning, and learning (see Fig. 1.3):

- Sensemaking refers to the fact that managers attempt to make sense of and interpret past situations and that their intuitive beliefs will also inform their future decisions.
- Resource allocation refers to the fact that firms invest in their business to maximize operating efficiencies and build competitive barriers.
- Strategic positioning refers to the fact that firms invest today to create opportunities for tomorrow and thereby "attempt to sustain performance across the unforeseeable future" (Bowman and Hurry 1993).
- Finally, learning refers to the acquisition of knowledge for the future (which should then drive strategy formulation).

This view of strategic themes, viewed through the option lens, bears a fair amount of similarity with the organizational cognition spiral (OCS) (Carayannis 1999) and its four stages of awareness and ignorance, even though the OCS is more analytical and dynamic, and outlines the process of higher order learning, while Fig. 1.3 is more descriptive.

Carayannis (1999) proposed the concept of the organizational cognitive spiral (OCS). The spiral consists of multiple levels of learning, each of which is built around four stages: "ignorance of ignorance, awareness of ignorance, awareness of awareness, and ignorance of awareness." The cycle starts with ignorance of ignorance (unknown unknowns), but eventually a need to know or curiosity triggers the move to the next phase. We now know that we are missing knowledge; we're at the awareness of ignorance stage. Once we have filled this gap, we are at the awareness of awareness stage. We know and we know that we know (tacit knowledge). Eventually, as we assimilate this knowledge, it will become part of who we are, and we will simply know, without being aware that we even know. This is the ignorance of awareness stage. This process first happens as described when we learn. However, it repeats itself for second-order learning (learning how to learn) and third-order learning (learning to learn how to learn) (Carayannis 1994, 1998, 2000).

Applying this framework to Bowman and Hurry's (1993) strategic themes can be interpreted as follows. At the sensemaking stage, managers are ignorant of their

Type of Analysis	Type of Decision Making: DELIBERATE (Content Themes)	Type of Decision Making: EMERGENT (Process Themes)
	Resource Allocation	**Sense Making**
Relatively Less Future Oriented	Organizations invest to maximize forecasted operating efficiency.	Organizational investment is the product if sense making, perceptual biases, and intuition.
	Strategic Positioning	**Learning**
Relatively More Future Oriented	Organizations invest to create new possibilities for future efficiency.	Organizational investment proceeds incrementally, as a result of accumulated learning.

Fig. 1.3 Strategy themes integrated by the option lens (Bowman and Hurry (1993) Fig. 2, p. 772)

ignorance (with respect to the framing of real options), and from their perception of events, they try to make some sense. Once they become aware of their ignorance, they don't yet view the organization's investments in terms of options, but they base the organization's resource allocation on the results of their previous sensemaking and intuition. As they enter the awareness of awareness stage, they start defining strategic choices in terms of real options. Finally, as the learning occurs and they incrementally build options based on organizational learning, they enter the ignorance of awareness stage, in which they almost "unconsciously" frame strategic choices in terms of options, as they assimilate this new option-identifying knowledge. Similarly to the organizational cognition spiral (OCS), this process of sensemaking→resource allocation→strategic positioning→learning repeats itself and builds upon the previous cycle. As the first cycle of the OCS deals with "learning" (L1) (Carayannis 1998, 2000), this first cycle contemplates the definition of organizational investments in terms of options. As the second OCS cycle deals with "learning to learn" (L2) (ibid), this second cycle contemplates framing organizational strategy and formulating strategic choices in terms of options. Finally, as the third OCS cycle deals with "learning to learn how to learn" (L3) (ibid), this cycle contemplates the framing of organizational strategy as options so as to enable further options as these unfold (see Fig. 1.4).

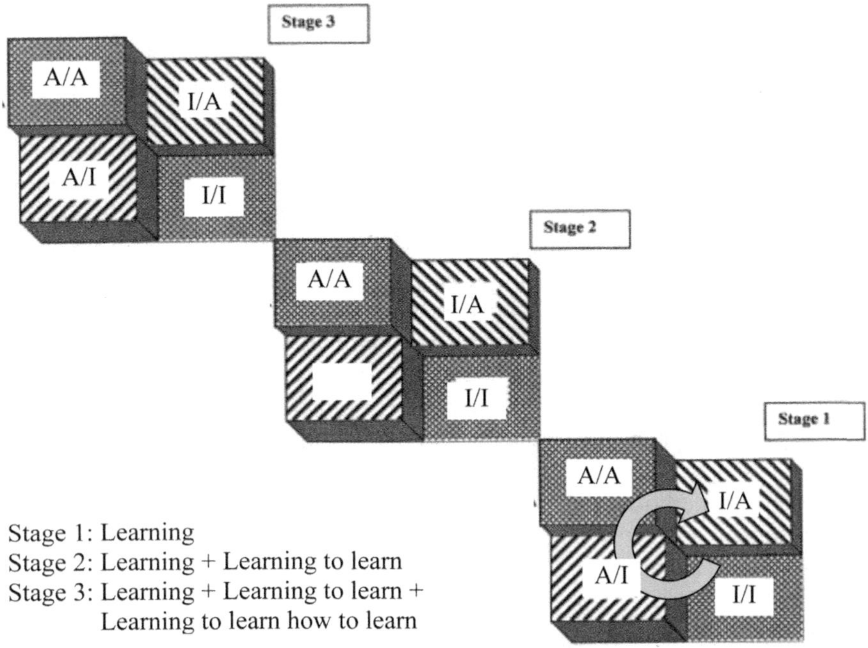

Source: adapted from Carayannis (1999), Figure 6, p. 229

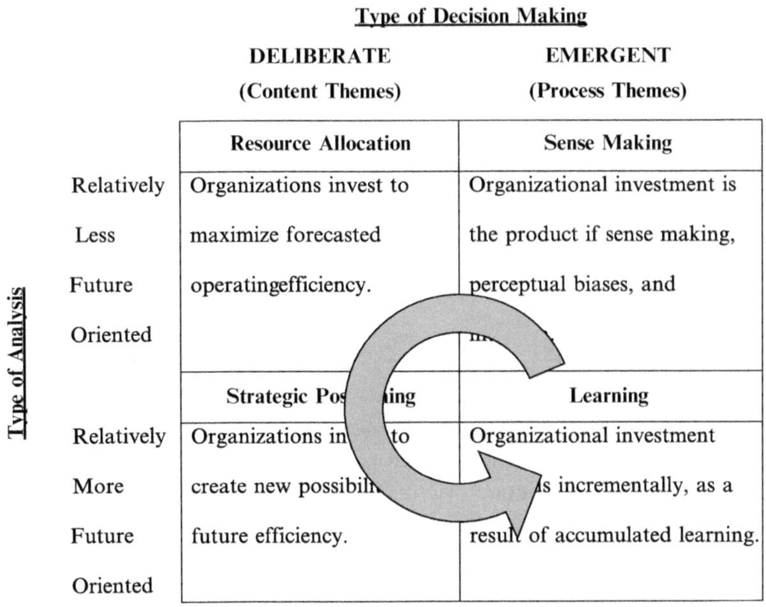

Fig. 1.4 Organizational cognition spiral (Adapted from Bowman and Hurry (1993), Fig. 2, p. 772; Carayannis (1999), Fig. 6, p. 229)

6 Summary of the Conceptual Framework

The proposed model explains how strategic decisions made during the life of a firm such as the creation of the firm, merger, acquisition, and closing of the firm take place. This is viewed through a real-options lens to highlight how co-opetition, coevolution, and co-specialization lead to the development of strategic knowledge arbitrage and serendipity, which becomes drivers of the real option about to be executed or not when the strategic decision is made.

The context for the proposed framework is one of the mode 3 system of knowledge production residing in Quadruple Innovation Helix fractal ecosystem architecture. This framework offers to apply embedded real options to heterogeneity and innovation events and processes. These elements drive firms to co-opete (*why*), coevolve (*when*), and co-specialize (*how*) in their environment. We then posit that firms that leverage strategic knowledge arbitrage and serendipity as drivers of the real options they define and manage through the formulation and implementation of their strategy will perform better over time.

This framework applies at various levels: micro (firm level), meso (industry or sector), and macro (regions or countries) (Porter 1990). One should also account for possible disparities according to the relative level of development of the countries in which the analysis takes place. Finally, as a continuum of disequilibria (Carayannis 1994) is central to the proposed framework, the model should also account for changes over time. As such, the model can be expressed as a function of the following six variables as fi(t, μ, m, M, ε, δ):

- Time (t)
- Microlevel of analysis (μ)
- Meso-level of analysis (m)
- Macro-level of analysis (M)
- Type of event (ε, e.g., merger, technology development, or acquisition)
- Relative stage of economic development (δ, e.g., low-, middle-, and high-income countries)

7 Empirical Findings

In addition to developing the conceptual framework described above, this work also sought to gather some empirical evidence of the model's validity. The empirical framework is "real-world practice oriented" and "concerned with 'what works'," which supports a pragmatic knowledge claim (Creswell 2003). Since the purposes of this work are descriptive and explanatory and we tried "to identify the existence or non-existence of a phenomenon and learn more on a little researched topic" (Creswell 2003), the chosen methodology was a two-phased mixed methods approach. The first phase encompassed two sequential aspects, one aspect helping

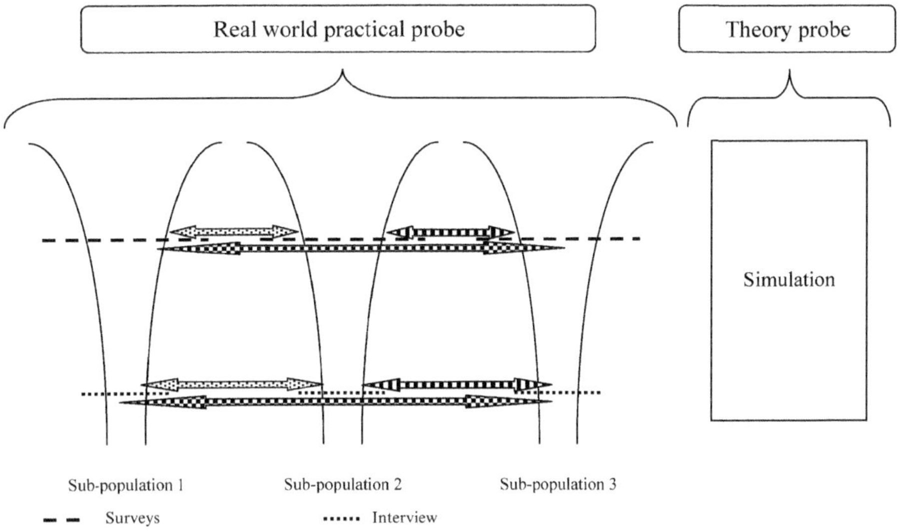

Fig. 1.5 Method visualization (Adapted from Carayannis, GWU MSTI Doctoral Seminar (2004))

inform the next and each one acted as a filter. A second and independent phase was used to probe the conceptual framework proposed earlier in this chapter.

Not any one method provided sufficiently comprehensive results. As such, we adopted a set of complementary, yet independent, methods to synthesize different perspectives (see Fig. 1.5):

- The survey (method 1a) provided real-world data to directly answer the research questions stated earlier.
- The interviews (method 1b) provided real-world data as an extension of the survey to assess the impact of key actors of the value chain (suppliers, clients, competitors, and regulators) on firms' technology choices.
- The simulation (method 3) aimed at providing modeling evidence to support the conceptual framework in which the research questions were formulated.

This work uses a multimethod approach to explore and illustrate, through all of these methodologies, how actors of innovation and entrepreneurship incorporate, implicitly or explicitly, real-options calculus in their choices, actions, and overall behaviors. In particular, it tries to illustrate the extent to which SKARSE™ (Carayannis 2008) governs the formulation and execution of real technology options.

The first stage of the work was survey based and compared and contrasted the technology investment decision-making process in the three subpopulations. The research purpose was to explore the differences in technology investment decision making between the three subpopulations. This was followed by seven interviews with the intent to learn more about the technology investment decision process of each organization and delve into more details on how the innovation and knowledge networks of the firms may impact their decision making.

This real-world-practice-oriented approach was complemented by an exploratory simulation of knowledge-based real-options execution. This is based on research published in the last year and allowed a first modeling look at the conceptual framework this dissertation offered. Methodological details about each phase are offered in Annexes 2, 3, and 4.

8 Contribution to the Literature

This book contributes to the literature both in terms of theory and empirical research. It contributes to the theory by building on previous work on the topic of strategy design and real options, in particular by Bowman and Hurry (1993) and by Carayannis (2008) and his research on strategic knowledge arbitrage and serendipity, co-opetition, coevolution, and co-specialization. This research argues that strategic knowledge arbitrage and serendipity allows companies to better define the content and exercise timing of their real options, as they take part in the co-opetition, coevolution, and co-specialization processes. In doing so, they leverage real options as risk management and uncertainty filters that will allow them to perform better over the longer term.

This book also contributes to the empirical research undertaken in the field of real options, little having been conducted on the actual use of real options in decision making (mostly, Bushy and Pitts 1997). This research provides exploratory data on the use of real options in decision making related to technology investments. Finally, it also provides a vehicle to test some of Carayannis' (2008) concepts related to co-opetition, coevolution, and co-specialization.

9 Book Structure

This book is organized into five chapters. This chapter introduced the topic of this work, its theoretical motivation, and briefly presented how the analysis is conducted. Chapter 2 presents a review of the main bodies of literature providing the context and content for this research, namely, decision-making theory, growth of the firm, innovation, options theory, innovation and knowledge clusters, and strategic knowledge and arbitrage. Chapter 3 presents the framework of analysis: the hypotheses to be tested, as well as the methodology and data that will be used to test these hypotheses. Chapter 4 presents the results of the work, while Chapter 5 presents the corresponding analysis and concludes the research.

Chapter 2
Literature Review

1 A Review of the Theory of Decision Making

The following sections review the main literature on the topics of choice and decision making. The main theories are reviewed first, followed by some considerations that may provide different contexts and explanations for them.

1.1 The Bounded Rationality Model

For Simon (1955) and March, Simon and Guetzkow (1958), reality is complex, but the human cognition is limited. Consequently, decision makers do not have exhaustive knowledge of all the possible alternatives, and they won't have all the necessary knowledge to adequately rank the alternatives and calculate their respective costs and benefits: rationality is bounded. On this basis, the decision maker does not attempt to find the optimal solution to a problem or the optimal choice for the organization; he will rather search a satisfying choice.

Simon (1979) also shows that often the goal of an organization cannot be connected directly to its operations. This will lead to the identification of a subordinate goal, what Simon calls "subgoal identification." Often, a subgoal is not unique, and its definition will be subject to the decision maker, his experience, knowledge, and the environment in which he evolves. And as complexity increases, decision making is often divided up among many specialists and their work coordinated through communication and authority relations.

C.M. Sipp and E.G. Carayannis, *Real Options and Strategic Technology Venturing: A New Paradigm in Decision Making*, SpringerBriefs in Business 31,
DOI 10.1007/978-1-4614-5814-2_2,

1.2 The Incrementalist View

Lindblom (1959) argued that means and ends are intertwined. Decision making is not a discrete event; rather, it becomes a step-by-step process, considering incremental changes from the status quo toward problem solving rather than decision making toward the achievement of a specific goal. For him, decision making is an evolutionary process, not a revolutionary one. The incrementalist view is also akin to "learning by doing" (E. Carayannis and Stokes 2000). It puts emphasis on the implementation of a decision rather than the analytical step that precedes the decision but focuses on continuous learning from the decision implemented and multiple feedback loops and decision adjustments, so as to reach the intended goal.

1.3 The Organizational Procedure View

For March (1978), decisions are viewed as the output of an organization's standard operating procedures rather than a specific and stand-alone decision-making process. In this context, Krabuanrat and Phelps (1998) highlighted the use and benefits of codified organizational experience so that decision-making input (whose understanding is common to everyone in the organization) can move through the standard operating procedures and lead to a decision. Codified organizational experience will ensure that everyone understands the rational and outcome of the decision-making process.

However, Das and Teng (1999) referred to the organizational procedure view as the "avoidance mode." They consider that a systematic decision-making process, such as standard operating procedures, can take away the thinking piece of decision making. They fear that in this type of environment, decision making is made at the cost of innovation.

1.4 The Garbage Can Model

Cohen et al. (1972) argued that the decision-making process is really a meeting point of multiple actors, multiple goals, and multiple views. They define the garbage can as "the meeting point of a problem in search of a solution, a solution in search of a problem and actors whose attention is divided, who come and go, but who happen to be at the meeting place of the problem and solution." As a critique, Mayer (2007) argued that the main drawback of the garbage can model is that since the decision point is really only a meeting point, it is impossible to adjust a decision made, and consequently, it is impossible to adjust the organization as a whole.

1.5 The Individual Differences Perspective

The individual differences perspective places the emphasis on individual decision making, rather than on organizational or group decision making. Keen and Morton (1978) showed that managers use different methods and come to different conclusions, not based on a rational choice but because of their personality, management style, and experience. As most decisions are perceived to emanate from a group decision-making process and a process that could be "standardized" for use within an organization, so as to have it move in one direction, this strings if research has not received as much attention as other decision-making theories and perspectives.

1.6 Naturalistic Decision-Making Perspective

Experience is the nexus of the naturalistic decision-making perspective. Klein (1999) argued that the decision-making process does not follow a rigorous method of solution identification, which is a key phase of the mainstream decision-making theories. Klein posited that the context in which the problem is posed is fundamental. Decision makers recognize a particular situation, its context, its possible outcomes, and the solution. As such, experience is key because the more experienced the decision maker, the easiest the recognition of the problem. Once the situation is recognized, the decision maker forms a scenario, which will then be tweaked until the decision maker is comfortable with it.

1.7 The Multiple Perspectives Approach

The multiple perspectives approach claims that all problems can be perceived from different angles. Mitroff and Linstone (1993) argued that no matter how rigorous the decision-making process, it will always be tainted by the perspective of the actors involved in the decision-making process. Thus, to reach a decision that is as comprehensive as possible and includes organizational, technical, and individual views, several stakeholders should be consulted.

1.8 The Effect of Speed on Decision Making

Kedia et al. (2002) claimed that traditional decision-making theories recommend that when facing uncertainty, decision makers should search for additional relevant information (Eisenhardt 1989; Milliken 1987; Simon 1987). By doing so, they delay the decision making and consequently alter actions and performance results.

Eisenhardt (1989) showed that the faster the decision-making process, the better. His work showed that the more successful companies analyze more relevant data, consider multiple alternatives simultaneously, and make faster decisions. Supporting his evidence, Baum and Wally (2003) also concluded that faster decision making has a positive impact on profitability and growth.

1.9 Conclusion

Decision-making theories can be broadly divided in to two schools of thoughts: the analytic and the experiential (incremental) schools of decision making. Both schools follow the three main steps in decision making: (1) problem definition, (2) identification, evaluation and selection of alternatives, and (3) implementation. The main difference between the two schools resides in how they weight the different steps. The analytic school focuses more heavily on the first two steps, and implementation is a simple execution of the carefully selected alternative, while the experiential schools focus more on the implementation of an alternative and feedback loops to adjust it as necessary.

2 Growth of the Firm

The following sections review the main theories of the growth of the firm – traditional but mostly endogenous (Paul Romer), new growth of the firm based on increasing returns.

2.1 The Resource-Based View of the Firm

Before Penrose, most of the literature considered the firm as an equilibrium model. Penrose (1959) proposed a view of the firm based on its productive resources, both physical and human. Penrose made clear that the resources are not themselves factors of productions, but that they rather produce a set of services. "The services yielded by the resources are a function of the way in which they are used – exactly the same resource when used for different purposes or in different ways and in combination with different types or amounts of other resources provides a different service or set of services. The important distinction between resources and services is not their relative durability; rather it lies in the fact that resources consist of a bundle of potential services [...]" (Penrose 1995, p. 25). The theoretical perspective of the firm as a learning organization confronts and combines the resource-based view (E. Penrose 1959) with recent approaches where the focus is on learning and the dynamic capabilities of the firm (Dosi et al. 1992).

Penrose (1959) not only redefined the firm, but she also identified elements of the growth of the firm, interested in knowing how much a firm can grow. She established that firms can grow either through diversification or through mergers and acquisitions. In each case, Penrose analyzed what the growth prospects were and concluded that after a period of growth, the firm would inevitably enter a period of decreasing returns.

2.2 Technology-Driven Firm Growth

Romer (1986) proposed a model of firm growth based on endogenous technological change. He adopted a neo-Schumpeterian growth model that focused on intentional and purposeful innovation. His main contribution was to move away from all the diminishing return models and offer a model in which growth rates can be increasing over time. For Romer, the long-term growth of the firm is primarily driven by the accumulation of knowledge. His model revolves around three elements: externalities, increase return on output production, and decreasing return on knowledge production.

2.3 Conclusion

When moving away from the traditional models of the firm, the two main theories are the resource-based view of the firm (Penrose) and the technology-driven firm (Romer). The main difference between the two theories is that Romer's model actually argued the existence of sustained growth rates for the firm.

3 Innovation

The following sections review the literature about innovation and how creativity, heterogeneity, and diversity can foster innovation. Here, we will understand innovation as "an invention successfully and sustainably brought to market" (E. G. Carayannis 2000–2009).

3.1 Incremental, Generational, Radical, and Architectural Innovation

The literature flourishes with a number of notions about innovations: whether they are incremental, generational, radical, architectural, continuous, discontinuous, and disruptive or not. Carayannis et al. (2003) proposed the following definitions of terms:

Table 2.1 Process/content/context/impact innovation framework

Process	Content	Context	Impact
Evolutionary innovation	Incremental innovation	Continuous innovation	Nondisruptive or disruptive innovation
Evolutionary innovation	Generational innovation	Continuous innovation	
Revolutionary innovation	Radical innovation	Discontinuous innovation	Nondisruptive or disruptive innovation
Revolutionary innovation	Architectural innovation	Discontinuous innovation	

- "Incremental innovations exploit the potential of established designs, and often reinforce the dominance of established firms. They improve the existing functional capabilities of a technology by means of small scale improvements in the technology's value-adding attributes such as performance, safety, quality and cost."
- "Generational or next-generation technology innovations are incremental innovations that lead to the creation of a new but not radically different system."
- "Radical innovations introduce new concepts that depart significantly from past practices and help create products or processes based on a different set of engineering or scientific principles and often open up entirely new markets and potential applications. They provide 'a brand-new functional capability which is a discontinuity in the ten-current technological capabilities'."
- "Architectural innovations serve to extend the radical-incremental classification of innovation and introduce the notion of changes in the way in which the components of a product or system are linked together."
- In evolutionary innovations, "technological change appears to follow a process of natural selection."
- In revolutionary innovations, "the change appears as a break or non-continuous change in the course of the technology."

Carayannis et al. (2003) also proposed to frame these various notions along four dimensions of process (of development, diffusion, and adoption), content (specific nature), context (environment), and impact (resulting socio-technological changes). Their proposed framework is shown in Table 2.1 and illustrated in Fig. 2.1.

Incremental innovations lead to incremental changes in the market (Christensen 1997; Tushman and O'Reilly 1996). Incremental innovations are based on learning how to better exploit exiting capabilities (Ireland et al. 2003). Although they don't lead to the exploitation of new entrepreneurial opportunities, incremental innovations are important "to help the firm derive maximum value from the firm's current capabilities" (Ireland et al. 2003).

Radical innovations lead to revolutionary changes in the market (Christensen 1997; Tushman and O'Reilly 1996). Disruptive innovations are based on exploiting new opportunities through a new combination of resources results in Ireland et al. (2003). Current market leaders often focus on improving their processes, as well as on the products and services they currently produce. They don't particularly feel

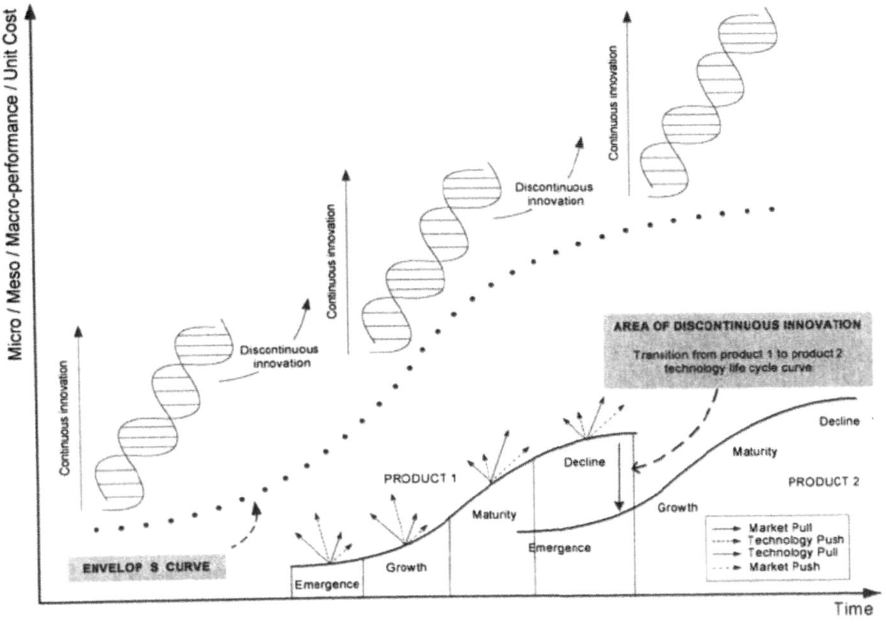

Fig. 2.1 Continuous and discontinuous innovation (Carayannis et al. 2003, p. 130)

comfortable changing the ingredients of a formula that works (Covin and Slevin 2002). Consequently, companies that perceive an unattended need or want out of that formula and can think of a way of responding to this demand or need in a way that implies a change in business model are usually not form the industry or not the market leaders (Christensen et al. 2002).

3.2 *Creativity and Innovation*

Koestler (1964) suggested that creativity, which subsequently leads to innovation, starts with a bisociation process. Bisociation is defined as "the simultaneous mental association of an idea or object with two fields ordinarily not regarded as related" (Merriam-Webster 2009). The greater the breadth of an individual's knowledge, the more likely the bisociation process (Ireland et al. 2003), but one can assume that a team whose members have diverse backgrounds would also be more likely to experience a bisociation process. This would explain why innovation is more typically the results of a team effort, while invention is more typically the result of an individual effort (or serendipity) (E. G. Carayannis 2000–2009). Bisociation is also the concept Smith and Di Gregorio (2002) built upon when they posited that creativity skills include the ability "to manage and manipulate diverse matrices of information, to suspend judgment as complexity increases, to consider broad categories of domain

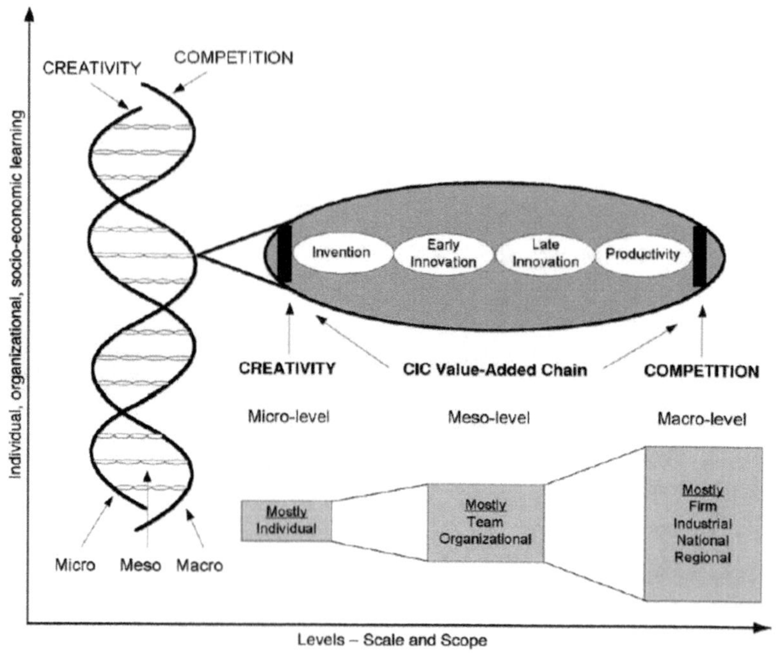

Fig. 2.2 The creativity-innovation-competition double helix (Carayannis and Gonzalez (2003), p. 594)

information, to remember accurately, and to recognize patterns or opportunities from alternative matrices of information." Bisociation and the type of knowledge one beholds can also be linked to different types of innovation. Should someone have deep knowledge of a subject, he would be more likely to develop incremental innovations, while someone with a broad knowledge across various fields is more likely to develop disruptive innovations (Ireland et al. 2003):

> Excellence is an art won by training and habituation. We do not act rightly because we have virtue or excellence, but we rather have those because we have acted rightly. We are what we repeatedly do. Excellence, then, is not an act but a habit. (Aristotle)

Taking Aristotle's quote in the context of bisociation, creativity is then not the result of serendipity but a process and a practice, so as to derive consistently innovation.

Carayannis and Gonzalez (2003) were inspired by the discovery of the double helix structure of the DNA and applied a similar structure to creativity, innovation, and competitiveness, in which creativity and innovation are part of a value-adding chain, starting with creativity, moving along early and late stages of innovation and improving productivity, to eventually result in increased competition (see Fig. 2.2). As we move up the value chain, we also move from a microlevel (individual) up to a macro-level environment for competition (firm, industry, or country). Carayannis and Gonzalez's empirical research showed that creativity and innovation are critical topics both for the private and the public sector. However, they went further and assigned a specific role to each actor of the GUI (government-university-industry). They posited that governments are responsible for ensuring a stable environment in which creativity and innovation can happen but also enforcing policies and rules that pertain to intellectual property protection, so as to provide not only a stable context but also one that fosters creativity and innovation. They encouraged universities to develop curricula that promote innovations, not just inventions, that is, inventions should be directed to answering business needs. Businesses should be responsible for developing and leveraging the resources necessary, that is, intellectual capital, social capital, and investments in research and development.

3.3 Heterogeneity, Diversity, and Innovation

Heterogeneity is defined as "the quality or state of being diverse and not comparable in kind" (Webster's Dictionary). Adam Smith (1776) understood that an increasingly sophisticated division of labor was the main source of productivity growth. He also knew that, when looking across tasks, it implied an increasing fragmentation of knowledge and corresponding heterogeneity. Carayannis (2009) proposed that heterogeneity be viewed as a cause and effect of innovation. As just mentioned, Smith (1776) showed that division of labor leads to knowledge heterogeneity. This heterogeneity, based on the principle of bisociation (Koestler 1964), will lead to innovation. Innovation itself, through the output generated by the innovation process, contributes to increased heterogeneity.

When analyzing dynamic organizations, McGrath and Boisot (2003) concluded that effective dynamic organizations had to be able to act in the absence of planned change. They deduced that this was shifting the job of strategists from focusing on a "variety-reducing planning process" to "promoters of conceptual diversity." Furthermore, March (1991) associated exploration in uncertain new areas with heterogeneity in resources accumulation, thus creating a potential for preferential access. The exploration of uncertain areas and flexibility about potential future choices lay at the heart of real options. This is why McGrath, Ferrier and Mendelow (2004) defined real options as "engines" of heterogeneity.

3.4 Conclusion

After a brief review of the basic forms of innovation, we focused on the role of creativity, diversity, and heterogeneity in innovation. The literature shows that diversity and heterogeneity contribute to innovation, which in turn contributes to diversity and heterogeneity, in a virtuous circle. A key point to take away from the perspective of this research is that real options are considered engines of heterogeneity and therefore fuel innovation as well.

4 A Review of Options Theory

The following sections review the main literature about options. After a brief introduction on financial options, the book summarizes the literature on real options, applied to discrete investment decisions, as well as corporate strategy.

4.1 Financial Options

Financial options are contracts between buyers and sellers that give buyers the right, but not the obligation, to buy (call option) or sell (put option) the underlying asset at a later date for a predetermined price. Should the market not be favorable to the buyers' "bets," buyers may decide to let the option expire, that is, not to exercise it. In 1973, Black and Scholes derived a differential equation that must be satisfied by the price of any derivative dependent on a nondividend paying stock. The Black-Scholes model is a rapid way of calculating the value of a European option.

For a call option,

$$C(S,t) = SN(d_1) - Ke^{-r(T-t)}N(d_2)$$

$$d_1 = \frac{\ln\left(\frac{S}{K}\right) + \left(r + \frac{\sigma^2}{2}\right)(T-t)}{\sigma\sqrt{T-t}}$$

$$d_2 = d_1 - \sigma\sqrt{T-t}$$

where:

- $N(\bullet)$ is the standard normal cumulative distribution function.
- $T-\text{t}$ is time to maturity.
- S is the spot price of the underlying asset.
- K is the strike price.
- r is the risk-free rate (annual rate, expressed in terms of continuous compounding).
- σ is the volatility in the log – returns of the underlying.

As simple as it is in this form, the model depends on very strict assumptions (Hull 2008):

- The stock price follows a geometric Brownian motion with μ and σ constant.
- The short selling of securities with full use of proceeds is permitted.
- There are no transaction costs or taxes.
- All securities are perfectly divisible.
- There are no dividends during the life of the derivative.
- There are no riskless arbitrage opportunities.
- Security trading is continuous.
- The risk-free rate of interest, r, is constant and the same for all maturities.

In 1979, Cox, Ross, and Rubinstein proposed the binomial tree pricing model in their article titled "Options pricing: a simplified approach." The greater the number of time steps in the tree, the closer the convergence toward the valuation of the Black-Scholes model. However, the more cumbersome the tree may be, it can be deemed more accurate than the Black-Scholes model in that it accommodates dividend-paying underlying assets as well as American-style options. Furthermore, the tree can be adjusted for more numerous outcomes at each step. Assumptions underpinning the binomial tree pricing model are otherwise similar to the ones of the Black-Scholes model.

For the purpose of this research, it is important to focus on two specific variables of option pricing: time to expiration and volatility. When either of these two variables increases, the value of the option (call or put) increases as well[1] (Hull 2008). Both variables are key elements defining uncertainty, which is typically considered as a negative element in investment decision making. This is precisely where real options attempt to bridge the gap between the finance theory and strategic planning.

4.2 *Overview of Real Options*

Keeping in mind the definition previously provided for financial options, a real option is similar but applied to a real, tangible, or intangible asset, rather than a financial underlying asset. A real option is the right, but not the obligation to undertake a business decision. Benaroch (2002) provided an extensive typology of real options: defer, stage (stop-resume), explore (pilot/prototype), alter scale (scale-up, i.e., expand, or scale-down, i.e., contract), abandon (switch use), outsource, lease, compound (two or more of the previously listed types), and growth (strategic).

Given the faster pace of change, tougher local and international competition, and more radical innovations and paradigm changes, business decisions face greater uncertainty. Thus, real-options analysis has become a more appealing approach for strategic decision making. The concept was first presented by Myers (1977) when he posited that some corporate assets, especially growth opportunities, could be viewed as call options: some investment opportunities grant the right, but not the obligation, to take specific operating action in the future or as Bowman and Hurry (1993, p. 761) explained it: "despite the absence of formal option contracts, they allow a similar pattern of investment behavior to occur."

Real-options analysis developed partly as an alternative that would overcome some shortcomings of the NPV analysis, in particular how inadequately it captures two significant sources of value for a company (Schwartz and Trigeorgis 2001 p. 47):

- "Operating flexibility, which enables management to make or revise decisions at a future time (e.g. options to defer, grow, or abandon the project) within a single project"

[1] Assuming early exercise is possible when considering time to expiration.

- "Strategic option value of a project with regard to future and follow-up investments"

Real-options identification is an important step. Not all managerial decisions are real options. In order to differentiate them, the existing literature has identified some key attributes (Kogut and Kulatilaka 1994; Dixit, Pindyck and Davis 1994; Copeland and Antikarov 2001; Durand et al. 2001):

- There needs to be a time frame associated with the option, so as to be able to compare on option with another.
- The decision has to be irreversible:
  - There should be a series of distinguishable and temporarily interdependent decisions (even if it's only two).
  - The irreversibility should be the same for all actors.
- Informational uncertainty is shared among all actors and that uncertainty diminishes as different sequences are realized (there has to be uncertainty about the underlying returns):
  - There is a series of probability events.
  - The decisions taken by the actors do not change the nature of the environment as it was defined at time t_0.
- An alternative strategic choice exists (management has to have flexibility to decide on the exercising of the option).

For Bowman and Hurry (1993, p. 762), "an option confers preferential access to an opportunity for investment choice – in the sense of gaining advantage over competitors, or in the sense of being better suited for one among several possible courses of action." "Real options represent choices (strategic or tactical) under conditions of risk and uncertainty about tangible and intangible (knowledge-based) assets (as compared and contrasted to financial assets) that encompass timing, selection and sequencing attributes of significance for the entity that may choose to exercise those options or not (individual, society or company)" (E. G. Carayannis 2009).

Option valuation differs from the valuation of other assets in that risk and uncertainty are key factors. Both volatility and time to expiration greatly impact the value of an option. It is how uncertainty is captured and included in the valuation model that has since then interested managers, since only some projects with very specific characteristics (e.g., asset-in-place investments) fit traditional and mainly linear valuation methodologies. Translating options into the physical world lead to the development of real options.

Contrary to standard valuation methodologies (e.g., discounted cash flows), and as supported by Day and Schoemaker (2000), real options are not static, they're dynamic. Real-options valuation is not a snapshot in time; rather the process needs to be managed. Day and Shoemaker showed that the process can be decomposed in four phases:

1. Adopting a real-options perspective
   Real options are not yet part of mainstream thinking. It therefore needs some adaptation to systematically look at business and recognize opportunities through

an option perspective. To some extent, this justifies current practice – all decisions can be seen as option decisions since all projects inherently leave room for managerial discretion in their implementation. To help adopting a real-options perspective, Bowman and Hurry (1993) argue that firms make use of what they call "sensemaking," that is, managers look at organizational actions taken and how resources were allocated in retrospect, trying to make sense out of them and retroactively identify what options were triggered. By this mean, managers can identify what Bowman and Hurry (1987) call "shadow options." To be able to strike strategic choices, managers must first be able to recognize them.

2. Creating and structuring real options
   Although project managers should be able to make the necessary decisions to ensure the success of their projects, it is important to think about this aspect beforehand and formally structure decisions to create future managerial flexibility. Moreover, most projects involve multiple or a sequence of decisions. Thus, one should look for opportunities to unbundle these decisions, so each represents one option. For the option to be most useful, all possible alternatives should be taken into account. One should therefore expand his consideration of additional possibilities for future action.
3. Valuing real options
   The valuation of real options integrates various aspects: financial returns but also strategic positioning and knowledge gained. The financial aspect probably receives the most attention. It can be assessed in various forms: using financial models (essentially Black-Scholes using a replicating portfolio), decision analysis (decision trees), or threshold assessment (combination of quantitative analysis and managerial judgment). It is in the valuation of real options that one probably best notices that real options are a systematic way of making explicit what could be considered a "gut feeling" on part of managers when they decide to go ahead with a project that has a negative net present value because additional value could be derived in form of knowledge or strategic positioning.
4. Implementing a real-options approach
   Because real options (and therefore their value) are dynamic, their implementation must be systematic. Assessing the value of the project only at one point in time defeats the superiority of the real-options approach; one needs to carefully monitor the progress of the projects, regularly test, and update his assumptions and based on this knowledge decide whether to exercise the option.

4.3 *Overview Real Options and Strategic Investments*

Since my main proposition is closely related to the fourth hypothesis stated by Bowman and Hurry (1993), we conducted a review of peer-reviewed articles and books that cited their 1993 article. During the second quarter of 2009, EBSCOHost database and Google Scholar yielded about 500 results in English, French, German, Spanish, and Portuguese that we reviewed to explore whether an empirical work as the one we are proposing had already been conducted (see Annex 1 for the bibliography reviewed). The result of this review was rather enlightening. A vast majority of articles research real-options theory or application at a disaggregated level, that is, applied to individual strategic decisions to be made. Of all the articles reviewed, only six contemplated strategy as a set of real options.

Tong, Reuer and Building (2004) provide evidence in support of looking at strategy as a portfolio of real options as their findings provide strong evidence that "certain corporate investments can provide future growth opportunities that are of value to the firm." But they also caution that this finding does not apply universally across investment type. They find a positive correlation between R&D investments and firm value attributable to growth options, but find that this does not hold true for investments in tangible capital.

Barnett (2003) argues that using real options may generate too much flexibility opportunities for the organization to handle, which may be detrimental. Barnett (2005) furthermore argues that applying real-options analysis requires undivided managers' attention, which is a scarce resource. Both of Barnett's articles warn against the overuse of real options to generate more opportunities that the organization and its management can handle and argue in favor of a correspondingly reasonably sized portfolio of real options.

Smit and Trigeorgis (2006) propose a framework for valuing and managing portfolios of real options, which if it is to be applied to the corporate strategy would assume that managers do indeed design and implement their corporate strategy as a portfolio of real options.

Finally, Kogut and Kulatilaka (2001) and Grewal and Tansuhaj (2001) analyze real options in the context of organizations defined as a set of capabilities, which relates directly to knowledge: capabilities can either be considered a particular kind of knowledge (Loasby 1998) or they can be viewed as collectively held knowledge (Spender 1996). My proposition analyzes strategy as a series of strategic investments that represent a portfolio of real options, and knowledge and learning, which are key firm capabilities, as drivers of the real options.

The following sections summarize current research on strategic decision making and real options, first for disaggregated investment decisions, followed by a section on corporate strategy as a portfolio of real options.

4.4 Real Options in Entrepreneurship

O'Brien et al. (2003) proposed an empirical test of the real-options logic to entrepreneurship. They argue that hardly any context would better fit the hypothesis for the definition of a real option than when an entrepreneur is considering launching a new venture. O'Brien et al. argued that at the very least, when faced with his entry decision, an entrepreneur would face the following choices: deferring, abandoning, switching inputs, altering scale, or staging the investment, all of which fit a real-options analysis framework. Their main findings reported that potential entrepreneurs are less likely to found a new business as uncertainty about industry conditions increase. They also found strong support for their prediction that this holds truer in industries that require investments that are largely irreversible. Consequently, entrepreneurs generally recognize the option to defer their investment. Even if they have to abandon one extra period of profit, they value the knowledge that they will get during that period to make a wiser decision.

Although these empirical results were in line with existing theory (Pindyck 1991; Dixit 1992; Dixit and Pindyck 1994) and some existing empirical evidence (Episcopos 1995), it came as a contradiction to propositions set forth by McGrath (1999) who argued that the high variance of outcomes in entrepreneurial activities will encourage it because the higher the variance in returns, the higher the value of the resulting option. It is also important that McGrath concludes by underscoring the importance of entrepreneurial failure because it is "easier to pinpoint why a failure has occurred than to explain a success" (McGrath 1999, p. 28). Thus, there is considerable value for entrepreneurs in learning from their and others' failures.

In addition to research from the entrepreneur's perspective, researchers have also studied real options and entrepreneurship through the lens of venture capital firms. Hurry et al. (1992) studied the strategic logic of Japanese and US high-technology venture capital firms. Their main hypothesis was that the investment logic of Japanese venture capital firms would reveal a shadow call option and would therefore make smaller individual investments, make a larger number of investments, follow longer-term strategic goals, and culminate the venture by retaining an option on the new technology. Their hypotheses in terms of number and magnitude of investments found strong empirical support. In addition, they found that Japanese venture capital firms tended to report more often that their venturing objective was technology, while US firms would report ROI. Furthermore, Japanese firms considered the possibility of future relations with the venture when selecting where to invest and typically retained ties to the venture, while US firms sold their stock or went to an initial public offering. These elements provide additional support to the view that Japanese high-technology venture capital firms pursue an option-based investment strategy.

Furthermore, the concept of entrepreneurship and its real-options analysis framework has also been linked to strategic entrepreneurship (Ireland et al. 2003). According to Ireland et al., strategic entrepreneurship generates superior firm performance through the simultaneous pursuit of multiple small entrepreneurial ventures within an established firm that can leverage its already-existing competitive advantages, which a small venture may lack, to improve the probability of success. Ireland et al. used the real-options logic to propose staged investment in multiple entrepreneurial ventures within the firm. For each smaller-scale-staged investment, the firm gains further knowledge about the venture and is in a better position to more efficiently allocate its resources across the portfolio of small ventures. With a well-balanced and well-managed option portfolio of entrepreneurial ventures, the firm should be able to allocate its resources to its most valuable ventures.

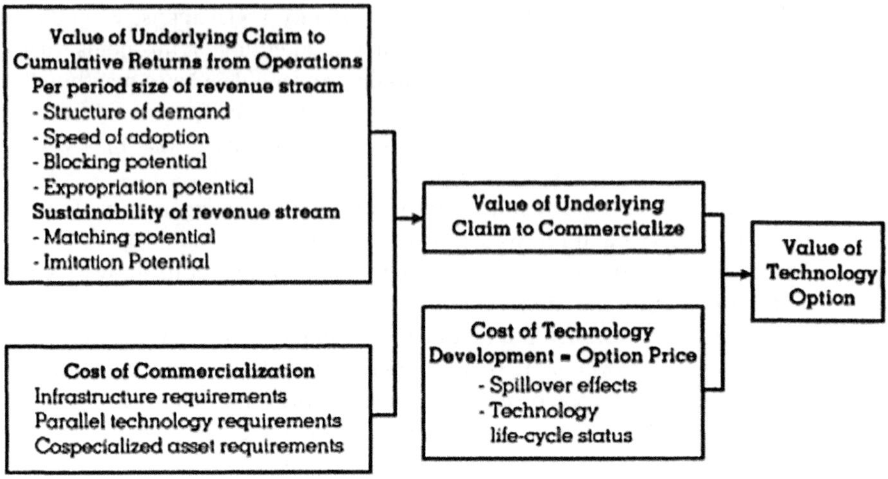

Fig. 2.3 Factors influencing the value of a technology positioning option (McGrath 1997 p. 988)

4.5 Real Options in Innovation and R&D

Technology investments typically require large investments and carry a large degree of uncertainty. As such, traditional valuation techniques may lead to substantial underinvestment. It is in this basis that McGrath (1997) proposed to further the real-options approach to technology positioning investments, integrating both technical and external uncertainties (see Fig. 2.3) and analyzing their impact on timing and option value. Technical uncertainty can be partially dealt with through-staged investments. Should investment not reduce that uncertainty, the firm could postpone its investment. When faced with external uncertainties, the firm can "take strategic actions to amplify the value of the option" (McGrath 1997). An important point that McGrath raises is that to be effective in raising the value of the option, existing resources endowment should reduce uncertainty for the firm while not reducing it for competitors.

Grenadier and Weiss (1997) proposed an option-pricing approach for investments in technological innovations that moves away from the discrete investment decision toward a sequence of interdependent investment decisions. In particular, they consider the innovation investment strategy as a sequence of embedded options. When deciding whether to invest in a new technology, a firm not only invests in the new technology but also holds an option on future updates (option on an option). When considering upgrading to a new technology, they hold an option to exchange the technology it currently has for the upgrade. Grenadier and Weiss identified four possible "migration strategies: (1) a *compulsive* strategy of purchasing every innovation, (2) a *leapfrog* strategy of skipping an early innovation, but adoption the next generation of innovation, (3) a *buy-and-hold* strategy of only purchasing an early innovation, and (4) a *laggard* strategy of waiting until a new generation of innovation arrives before purchasing the previous innovation." Their model contemplated the effect of speed of innovation and expected technological growth. They concluded that in environments that sustain a rapid pace of innovation, leapfrog and laggard strategies are the most likely, while compulsive and buy-and-hold are more likely to take place in a slow-paced environment. Leapfrog and compulsive strategies are more likely to take place when the expected innovation growth increases, while the likelihood of laggard strategy decreases, and the one of the buy-and-hold remains somewhat constant.

As mentioned earlier, Hurry et al. (1992) researched the strategic investment logic of Japanese high-technology venture capital firms. Not only did their research show that these firms approached the venture capital process as a real option, but their findings regarding the culmination of the venture investments were also very telling regarding how Japanese firms view the acquisition of innovations and technology. While the venture capital process is managed as a real option, it is also part of another real option to acquire technology. This became apparent in the Hurry, Miller, and Bowman study when the fact that US venture capital firms exited their projects by profit taking was contrasted to the exit of the Japanese companies, which listed retaining stock in the venture or transferring it to a sister firm in their keiretsu or joint ventures. This was consistent with the objectives of the venture financing the firms disclosed, such as "new technology and business development," "strategic move to new areas," "proprietary technology," or "diversification." In addition, the Japanese firms invested smaller amounts in the new ventures than comparable US firms, even though the full development and deployment of the technology would require much larger investments, which suggests that their involvement in the venture capital process was part of a staging option strategy to learn more about the technology and maybe pilot it before fully engaging.

4.6 Real Options in the International Development of Firms

Kogut (1991) proposed to analyze joint ventures in a real-options framework. He hypothesized that "the timing of the acquisition should be triggered by a product market signal indicating an increase in the venture's valuation." Using a sample of 92 manufacturing companies, he developed empirical results that support his hypothesis, showing that "unexpected growth in the product market increases the likelihood of acquisition." One of Kogut's main contributions is to propose something that goes against the mainstream literature in organizational behavior. Most of the literature has been focusing on minimizing risks, while Kogut is suggesting that uncertainty can also benefit firms if they use a more flexible production process or organizational design.

Entering a new market is always perceived as high risk, and such an investment requires a lot of attention, especially concerns about having to control the operations over a great geographic distance and operating in a foreign culture (Hymer 1960). Kogut and Kulatilaka (1994) proposed to look at a firm's multinational network through a real-options lens and highlight the operational flexibility that such a network provides as a valuable option. They used dynamic programming to compute a numerical example, which by far supports their theoretical model. Their results display an incremental profitability rarely short of 10 % and up to 35 %, which is far from negligible. Although the real options framework can be applied to many strategic investment decisions, they believe that it is particularly suited to the internationalization of a firm. They conclude that "despite the popular notion of the riskiness of international markets, it is [this] *uncertainty* that drives the opportunities available to the firm that is multinational, as opposed to only domestic, in its investments and operations," (emphasis added) which lays at the heart of real-options theory.

4.7 Real Options in Merger and Acquisitions and Restructuration of Firms

Hurry (1993, p. 70) defines global restructuring as "the strategic process of striking options to capture complementary opportunities, across global firms, for changes on portfolio scope and financing." Hurry based his approach on Bowman and Hurry's (1987) view of the value of a firm, in that it can be analyzed as the value of the business assets, but in addition, also the value of opportunities created by these assets, which represent a portfolio of options (of calls to expand the firm and of puts to contract the firm). Striking some of these options leads to the reconfiguration of the company's portfolio. One of the main contributions of Hurry related to strategic management is that his theory advocates the conceptualization of strategy as a dynamic process. He posits that this would be particularly useful in some areas of research where results are debatable: citing Trautwein (1990), "areas such as merger and acquisitions, where cross-sectional approaches have not produced unequivocal results."

Mergers and acquisitions (M&A) have faced serious critique in the literature, most scholars arguing that the acquiring firm reaps no or little benefits (Jensen, Ruback, Field and Park 1983; Loughran and Vijh 1997). Smith and Triantis (1995) posited that more gains are possible when analyzing such investment decisions as part of a framework that includes growth options, flexibility options (strategic diversification rather than purely financial), and divestiture options. Growth can be achieved through strategic acquisition, but Smith and Triantis highlight that growth options exist both for the acquiring as well as the target firm. The target firm may need larger infrastructure and access to cheaper capital, which the acquiring firm can provide, and the acquiring firm may benefit from knowledge transfer or access to new markets and collaborators. As with any option framework, Smith and Triantis highlighted the necessity of adequately identifying the types of options that exist in a particular acquisition transaction but also carefully developing and exercising them, so as to reach the full benefits of the synergies between the target and the acquiring firm.

The careful development and exercise of real options also lies at the heart of Carayannis' (2008) theory of strategic knowledge arbitrage and serendipity as real-options drivers. Carayannis (ibid) stated that "strategic knowledge serendipity provides the terms of the option, while strategic knowledge arbitrage influences the decision making." Mergers and acquisitions illustrate this point very well, since they involve recombining and reallocating assets, both tangible and intangible. M&A learning at the tactical, operational, and strategic levels (Carayannis 1994, 1998, 1999, 2000) comes together to form the base of the M&A strategic knowledge arbitrage and serendipity. This, in turn, will define the terms of the options created by the newly merged company and inform exercise timing (or not), which, if exercised carefully will contribute to improved post merger integration and merged firm performance.

4.8 *Organizational Strategy and Real Options*

Luehrman (1998) showed that business strategies can be seen as a portfolio of real options. As Luehrman explains it, "a strategy defines the path a company wants to take from where it currently stands to where it wants to be." This involves choices, learning, and sequence of major decisions. Carayannis (E. G. Carayannis 2000–2009) defines strategy as "the competence to discern the non-obvious faster, better, cheaper, longer." Analyzing projects from a real-options perspective not only forces managers to think several moves ahead (which is nothing new to them), but the option pricing quantifies the value of follow-on opportunities better than standard discounted cash flow.

Bowman and Hurry (1993) also supported the view of an organizational strategy as a bundle or portfolio of options. In addition, they prolonged their theory to impact organizational structure. They based their proposition on Merton's Theorem 7 (1973, p. 148):

Theorem 7. If $S_i = S_j = S, i, j = 1, 2, \ldots, n;$

$$Z_{n+1}(\tau) \equiv \Sigma_1^n \lambda_i Z_i(\tau)$$

for $\lambda_i \in [0,1]$ and $\Sigma_1^n \lambda_i = 1$, then

$$F_{n+1}(S, \tau; E) \leq \Sigma_1^n \lambda_i F_i(S, \tau; E).$$

Merton's Theorem 7 states that generally diversification hurts an option on a portfolio, as it is more valuable to hold a portfolio of options than to have an option on a portfolio. As such, an organization will be better off if its structure allows it the flexibility to strike the various options it holds when appropriate rather than having a single strike power at the top of the organization that would exercise this right by impacting a lower level of the organization that in its turn would contain a bundle of options (therefore, exercising an option on a portfolio of options rather than managing a portfolio of options).

This real-options-based approach to conceiving strategy is congruent with another perspective on strategy as "the competence to discern the non-obvious faster, more clearly and more accurately" (E. G. Carayannis 2000–2009), and it further highlights the role and influence of SKARSE™ and learning (E. G. Carayannis 2008). Luehrman (1998) also recommends that for project valuation, option-pricing techniques should be used as complements to existing techniques, not as replacements. In doing so, he emphasizes the creative aspect of finance instead of the more traditional due diligence. If used early on and given adequate freedom, finance can be more creative and play a more important part in the strategy design, that is, it can "contribute insightful interpretive analyses of sequences of decisions that are purely hypothetical-that is, while they are still mere possibilities." Gaining this knowledge and knowing how, why, and when to use it is akin to a knowledge-based real option. It requires a mind shift that allows managers to consistently identify real options in their business and proactively manage them. Knowledge serendipity provides them with the necessary knowledge to define the terms of the real options, and knowledge arbitrage allows them to decide if, when, and how best to execute them. The creation and growth of a new venture can be seen as a portfolio of real options, whether embedded, serial, or parallel. This portfolio of real options can be envisioned in a mode 3 context (multi-nodal, multimodal, multilayer), in which real options could be multi-nodal depending on where the entity is situated on the path of real options, multimodal depending on which kind of real options is considered (embedded, serial, parallel), and multilayer depending on the level of synthesis one is considering (firm, industry, national levels) (E. G. Carayannis 2009).

5 Innovation Networks and Knowledge Clusters

The following sections review the literature on innovation networks and knowledge clusters. We will consider the following definition of terms per Carayannis (2009):

> *Innovation Networks* are real and virtual infra-structures and infra-technologies that serve to nurture creativity, trigger invention and catalyze innovation in a public and/or private domain context (for instance, Government-University-Industry Public-Private Research and Technology Development Co-opetitive Partnerships).
>
> *Knowledge Clusters* are agglomerations of co-specialized, mutually complementary and reinforcing knowledge assets in the form of "knowledge stocks" and "knowledge flows" that exhibit self-organizing, learning-driven, dynamically adaptive competences and trends in the context of an open systems perspective.

5.1 Knowledge Creation Within the Firm

Learning within firms can take place many different ways (Simon 1991). Although the form can be very varied, Fuchs (2001) established that knowledge creation within firms is closely related to ongoing activities. Lawson and Lorenz (1999) expanded this idea and ascertained that what each employee learns is somewhat connected to what other employees may know or learn. However, each area of expertise develops its own body of knowledge, and when a firm develops too many

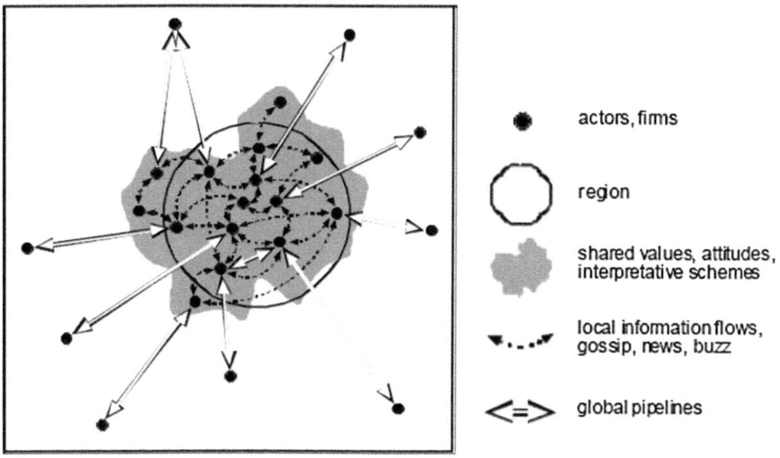

Fig. 2.4 Local buzz and global pipelines

priorities, with their corresponding specialized knowledge, the common knowledge base of the firm significantly diminishes. Nonaka and Takeuchi (1995) showed that cross division knowledge creation therefore results in lower production costs. The result of intra-firm knowledge creation is mostly incremental learning.

5.2 *Knowledge Creation Across Firms*

Contrary to incremental intra-firm learning, the literature has contemplated for quite some time the fact that radical knowledge creation is usually an interactive process across several firms (Rosenberg 1982; Freeman 1985; Kline and Rosenberg 1986; Lundvall 1988). The division of labor and increased specialization have led to a growth of knowledge and of the economy (Adam Smith 1776; Young 1928). However, in a case similar to a firm too diversified, if the knowledge base across firms is too diverse, the interaction across firms and interfirm projects will be more difficult to manage (Storper and Venables 2004).

5.3 *Knowledge Creation Across Clusters*

The literature about clusters seems to be convinced of the theoretical superiority of geographically close clusters, but there is little empirical evidence to support this statement (Bathelt 2001). Owen-Smith and Powell (2002) showed that being local is not enough for knowledge transfer. Knowledge also travels through innovation and social networks, which are not defined geographically (see Fig. 2.4).

In support of this, Uzzi (1996, 1997) also demonstrated that "over-embeddedness" was not a positive factor: there is a decreasing return on proximity. Burt (1995) also emphasized the importance of having ties with remote networks to bridge structural holes. For a firm to enhance its "absorptive capacity," it is important to have heterogeneous knowledge; otherwise it won't be able to fully leverage the knowledge transmitted (W. M. Cohen and Levinthal 1990).

5.4 *Toward a Mode 3 Ecosystem*

Carayannis (2009) has built on the existing literature to develop a mode 3 of knowledge-generating system that is based on innovation networks and knowledge clusters. This mode 3 system possesses the following attributes:

- Multilayered (e.g., it happens a different levels: world/country/sector/firm)
- Multimodal (i.e., based on mode 1 and mode 2 system of knowledge production)
- Multi-nodal (i.e., there are multiple connection points within a layer: high connectivity)
- Multilateral (public, private, university, civil society – i.e., the four strands of the Quadruple Innovation Helix)

The four strands of the quadruple helix are government, universities, industry, and civil society. Adding civil society to the triple helix (GUI) combines bottom-up initiatives and practices with top-down policies. In essence, civil society is not only a fourth strand but the bridge and connector between the three remaining strands.

The quadruple helix is an architectural blueprint of the innovation ecosystem, how different sectors are intertwined. The triple helix was meant to work with modes 1 and/or 2. Now, the Quadruple Innovation Helix adds a fourth strand (civil society) and increases connectivity within and across strands. The Quadruple Innovation Helix is meant to work in mode 3 and will not work in modes 1 and 2 because of the diversity and heterogeneity of the participants (e.g., big companies and government interacting with individuals: this interaction is only possible through the higher interconnectivity of mode 3) (E. G. Carayannis 2009).

5.5 *Conclusion*

In essence, innovation networks and knowledge clusters are meta-options. A firm has the option to network or not, to cluster or not. When looking at a system that consists of innovation networks and knowledge clusters, one can see a bundle of embedded options.

6 Strategic Knowledge Arbitrage and Serendipity (SKARSE™)

The following sections review the literature on strategic knowledge arbitrage and serendipity and on how these concepts become real-options drivers. As the concepts are still fairly young (and there is little in the literature), we will start with a definition of terms.

6.1 Definition of Terms

As the literature on the role of knowledge in gaining and sustaining competitive advantage became more active, more studies were conducted about how to keep abreast of one's competitors' development and concluded that much of this knowledge was involuntarily diffused, mostly by third parties (e.g., investment bankers, board members, consultants) (Semadeni 2001). The early definition of knowledge arbitrage (both in academia, Kao 1996, and in the press, *McKinsey Quarterly* 1998) was related to multinational issues: "with individuals who span national and cultural boundaries capitalizing in the arbitrage of knowledge from one country or culture to another" (Semadeni). Semadeni focused his research on the role of management consultants, which he perceived as dual; the role of knowledge arbitrageur, in which the consultant transmits knowledge; and the role of knowledge arbiter, in which there is much room for discretion regarding what knowledge is subject to arbitrage or not.

Davis (2000) highlighted the importance of defining knowledge patterns, especially in the knowledge-based economy for which knowledge is a fundamental input, but also a factor that is fast changing and growing increasingly complex. He posited that pattern recognition "supports effective and rapid decision making." Davis identified 29 knowledge patterns, including knowledge arbitrage, which he defined as "the efficient sourcing and distribution of ideas and products drawing on the best ideas and lowest priced inputs from around the globe." However, Davis' study did not go into the details of how to leverage knowledge arbitrage.

Carayannis (2008) proposed a definition that specifically applies to strategic knowledge:

> Strategic knowledge arbitrage. This refers to the ability to distribute and use specific knowledge for applications other than the intended topic area. More specifically, it refers to the capacity to create, identify, reallocate and recombine knowledge assets more effectively and efficiently to derive, develop and capture nonappropriable, defensible, sustainable and scalable pecuniary benefits.

As a complement to strategic knowledge arbitrage, Carayannis also defined strategic knowledge serendipity as follows, coining the concept of SKARSE™ (strategic knowledge serendipity and arbitrage):

Strategic knowledge serendipity. This term refers to the unintended benefits of enabling knowledge to 'spill over' between employees, groups and functional domains ('happy accidents' in learning). More specifically, it describes the capacity to identify, recognize, access and integrate knowledge assets more effectively and efficiently to derive, develop and capture non-appropriable, defensible, sustainable and scalable pecuniary benefits.

6.2 Leveraging SKARSE™ for Real Options

As described by Carayannis (2008), firms move through the stages of co-opetition, coevolution, and co-specialization (C3) by developing and exercising series of real options. To achieve robust competitiveness and sustainable entrepreneurship, firms have to develop an "unfair" competitive advantage (even for a short period of time). They can develop that competitive advantage by exercising real options. But to increase the probability that this competitive would be "unfair" for at least a period of time, firms should leverage strategic serendipity and arbitrage and use them as drivers of the real options they define and manage. "Strategic knowledge serendipity provides the terms of the option, while strategic knowledge arbitrage influences the decision making" (Carayannis 2008, p. 349).

6.3 Conclusion

The literature on strategic knowledge and serendipity is still in its infancy, especially in terms of recommendations on how to leverage it to derive a sustainable competitive advantage. The most advanced one is the framework proposed by Carayannis (2008), building on C3 and leveraging SKARSE™ as real-options drivers.

7 Conclusion

This chapter covers the main elements of the conceptual framework proposed in Chap. 1, namely, real options, innovation, innovation networks and knowledge clusters, and strategic knowledge serendipity and arbitrage. It has also reviewed the literature for the concepts that provide the context for the conceptual framework: decision making and the growth of the firm.

The key take-away of this chapter appeared under the real-options literature review, the main conclusion being that even though real-options analysis has become increasingly popular, it still tends to focus on decision making for discrete investments. Bowman and Hurry (1993) and Luehrman (1998) are the key researchers that have written about real-options frameworks applied to corporate strategy. This provided the opportunity for the work presented here forth.

Chapter 3
Findings

1 Survey Analysis

The following analysis is based on 61 completed surveys (92.42% completion rate as 66 surveys were viewed and started).

The respondents were to be classified into three subpopulations (technology-driven, TD; technology-based, TB; and technology-neutral, TN). TD firms could be identified from NAICS codes, but TN and TN could not be accurately identified before launching the survey. To identify which respondent corresponded to which subpopulation, we used a filter (self-identification) question (see question 2 of the survey in Annex 2). The answers to this question were used for poststratification. Poststratification is defined as an "approach is typically implemented due to a lack of prior knowledge of an appropriate stratifying variable or when the experimenter lacks the necessary information to create a stratifying variable during the sampling phase" (Pedhazur and Schmelkin 1991).

1.1 Overview of the Survey Respondents

Of the 61 respondents, 24 belonged to TD, 31 to TB, 2 to TN, and 4 refused to answer. Given the underrepresentation of TN, from this point forward, all the survey analysis is based on the TD and TB responses. TB was generally more inclined toward low-risk projects and a balance of low- and high-risk projects. TD was more inclined toward companies with a greater number of employees. Information technology services were the most represented industry, but with only 13% of all respondents, respondents were spread across various industries. 89% of all respondents' company was headquartered in the USA, mostly in Washington D.C. Metropolitan Area. See Annex 5 for graphical illustrations of TD and TB profiles.

C.M. Sipp and E.G. Carayannis, *Real Options and Strategic Technology Venturing: A New Paradigm in Decision Making*, SpringerBriefs in Business 31,
DOI 10.1007/978-1-4614-5814-2_3,

1.2 Hypotheses Analysis

This research envisioned comparing and contrasting the results gathered across three subpopulations. Given the respondents' profile (only two companies in TN), we were only able to compare results between the technology-driven and technology-based subpopulations (TD and TB).

It should be noted that the Likert scale used in the survey was coded as follows, with "1" corresponding to a higher degree of importance and "5" a lower degree of importance:

- Very important, "1"
- Important, "2"
- Neutral, "3"
- Unimportant, "4"
- Not important at all, "5"
- N/A, "6"

1.2.1 Primary Hypothesis

H_0: A real-options approach drives strategic decision making in technology venturing.

1.2.2 Secondary Hypotheses

H_1: The use of real options in decision making for technology investments increases in a co-opetitive environment.

H_{1A}: The use of real options in decision making for technology investments *is higher in TD than in TB* in a co-opetitive environment.

H_1 expected results: The use of real options in decision making to increase in a co-opetitive environment.

We used the Wilcoxon rank-sum (Mann–Whitney) test on the following null hypothesis:

Null hypothesis:
The use of real options in decision making for technology investments is equal in TD and in TB in a co-opetitive environment.

The use of real options was operationalized through the perceived importance of the following attributes of flexibility: abandonment, postponement, rescaling, and growth. The data for these attributes was also aggregated into "flexibility." The null hypothesis was tested for each of these attributes. See Annex 6 for the results of all the tests.

As shown in Table 3.1 below, the results suggest that there is a statistically significant difference between the underlying distributions in the cases marked with

Table 3.1 Statistical difference between TD and TB for options in co-opetition

	C1Postpone	C1Abandon	C1Rescale	C1Growth	C1Flex
p value	0.1844	0.0386*	0.1055	0.0011*	0.0149*
	C1Postpone 2008	C1Abandon 2008	C1Rescale 2008	C1Growth 2008	C1Flex 2008
p value	0.6426	0.5428	0.0794**	0.0492*	0.4203

*Significant at $\alpha=5\%$, **significant at $\alpha=10\%$

Table 3.2 Statistical difference between TD and TB for options in strategic coevolution

	C2SPostpone	C2SAbandon	C2SRescale	C2SGrowth	C2SFlex
p value	0.4627	0.6617	0.0939**	0.0034*	0.1652
	C2SPostpone 2008	C2SAbandon 2008	C2SRescale 2008	C2SGrowth 2008	C2SFlex 2008
p value	0.4844	0.872	0.5903	0.0064*	0.4006

*Significant at $\alpha=5\%$, **significant at $\alpha=10\%$

an asterisk. The null hypothesis can be rejected for the option to abandon, to grow and flexibility today, and for the option to rescale and grow in 2008. Which group has the higher rank by looking at how the actual rank sums compare to the expected rank sums under the null hypothesis? TD has a lower rank, showing a greater degree of importance attributed to the options presented.

H_2: The use of real options in decision making for *strategic and nonstrategic* technology investments increases with risk and uncertainty.

H_{2A}: The use of real options in decision making for *strategic* technology investments increases with risk and uncertainty.

H_{2A1}: The use of real options in decision making for *strategic* technology investments is higher in *TD than in TB* with risk and uncertainty.

H_{2A} expected results: The use of real options in decision making for strategic technology investment to increase with risk and uncertainty.

We used the Wilcoxon rank-sum (Mann–Whitney) test on the following null hypothesis:

Null hypothesis:
The use of real options in decision making for strategic technology investments is equal in TD and in TB with risk and uncertainty.

As shown in Table 3.2 below, the results suggest that there is a statistically significant difference between the underlying distributions in the cases marked with an asterisk. The null hypothesis can be rejected for the option to rescale and grow today and for the option to grow in 2008. Which group has the higher rank by looking at how the actual rank sums compare to the expected rank sums under the null hypothesis? TD has a lower rank, showing a greater degree of importance attributed to the options presented.

Table 3.3 Statistical difference between TD and TB for options in nonstrategic coevolution

	C2Postpone	C2Abandon	C2Rescale	C2Growth	C2Flex
p value	0.2691	0.748	0.9388	0.3802	0.6633
	C2Postpone 2008	C2Abandon 2008	C2Rescale 2008	C2Growth 2008	C2Flex 2008
p value	0.5488	0.3843	0.9172	0.3656	0.8025

*Significant at $\alpha=5\%$, significant at $\alpha=10\%$

H_{2B}: The use of real options in decision making for *nonstrategic* technology investments increases with risk and uncertainty.

H_{2B1}: The use of real options in decision making for *nonstrategic* technology investments is higher in *TD than in TB* with risk and uncertainty.

H_{2B} expected results: The use of real options in decision making for nonstrategic technology investment to increase with risk and uncertainty.

We used the Wilcoxon rank-sum (Mann–Whitney) test on the following null hypothesis:

Null hypothesis:
The use of real options in decision making for nonstrategic technology investments is equal in TD and in TB with risk and uncertainty.

As shown in Table 3.3 below, the results suggest that there is no statistically significant difference between the underlying distributions in any of the cases presented. The null hypothesis cannot be rejected for any options for nonstrategic investments, neither today nor in 2008.

H_{2C}: The use of real options in decision making applies *more often for strategic than nonstrategic* technology investments in *TD and TB*.

H_{2C} expected results: The use of real options in decision making for strategic technology investment to be more frequent than for nonstrategic investments.

Null hypothesis:
The use of real options in decision making is the same for strategic than nonstrategic technology investments in TD and TB.

To test for this hypothesis, the data had to be manipulated. We created a variable "flex" appending the flexibility data from the nonstrategic investment question to the data of the strategic investment question. Each of these data point was matched to a "strat" variable that took a value of one when the data point was from the strategic question and 0 when the data point was from the nonstrategic question. This "strat" variable became the grouping variable so as to compare ranks for strat=1 data points to ranks for start=0 data points.

Since this hypothesis testing involved testing within the same sample and not two independent samples, we did not use the Wilcoxon rank sum (Mann–Whitney) but rather the Kruskal-Wallis test (nonparametric analogue to the ANOVA).

Table 3.4 Statistical difference in option use between strategic and nonstrategic investments for TD

strat	Obs	Rank sum
0	24	689.50
1	24	486.50

Chi-squared=4.380 with 1 d.f.
Probability=0.0364
Chi-squared with ties=4.436 with 1 d.f.
Probability=0.0352

Table 3.5 Statistical difference between TD and TB for options in co-specialization

	C3Postpone	C3Abandon	C3Rescale	C3Growth	C3Flex
p value	0.2892	0.7681	0.0058*	0.0004*	0.0514**
	C3Postpone 2008	C3Abandon 2008	C3Rescale 2008	C3Growth 2008	C3Flex 2008
p value	0.1131	0.5188	0.0016*	0.0037*	0.0283*

*Significant at $\alpha=5\%$, **significant at $\alpha=10\%$

As shown in Table 3.4 below, the difference in sum of ranks of flexibility is statistically significant between strategic and nonstrategic technology investments in TD, leading to the rejection of the null hypothesis. The same test run for TB showed no statistical significance, and we, thus, cannot reject the null hypothesis for TB.

H_3: The use of real options in decision making increases when the technology investment under consideration leads to new business areas (product, services, and/or processes).

H_{3A}: The use of real options in decision making *is higher in TD than in TB* when the technology investment under consideration leads to new business areas (product, services, and/or processes).

H_3 expected results: The use of real options in decision making to increase when investments under consideration lead to new business areas.

We used the Wilcoxon rank-sum (Mann–Whitney) test on the following null hypothesis:

Null hypothesis:
The use of real options in decision making is equal in TD than in TB when the technology investment under consideration leads to new business areas (product, services, and/or processes).

As shown in Table 3.5 below, the results suggest that there is a statistically significant difference between the underlying distributions in the cases marked with an asterisk. The null hypothesis can be rejected for the option to rescale and to grow and flexibility both today and in 2008. Which group has the higher rank by looking at how the actual rank sums compare to the expected rank sums under the null hypothesis? TD has a lower rank, showing a greater degree of importance attributed to the options presented.

Table 3.6 Characteristics of interview participants

Interview #	Industry	Size	Location
1	Aviation and aerospace	More than 5,000	n/a[a]
2	Financial services	1,001–5,000	USA, DC
3	Financial services	501–1,000	USA, DC
4	Computer software	Less than 50	USA, TX
5	Newspapers	501–1,000	UK
6	Renewable and environment	Less than 50	USA, MA
7	Banking	More than 5,000	USA, CA

Interview #	Suppliers' location	Clients' location	Competitors' location	Regulators' location
1	International	International	International	International
2	Local, regional	International	International	International
3	National	National	National, international	National, international
4	National	National	National, international	National
5	International	Print: 90% national online: 1/3 national, 2/3 international	Print: national online: international	National
6	International	n/a[b]	National, international	National, international
7	International	Local, regional, national, international	Local, regional, national, international	National, international

[a]Given the few numbers of actors in this industry, the location of the company is not to be disclosed
[b]This firm is in its infancy and does not have clients yet

2 Interview Analysis

From the 61 survey respondents, 16 indicated their contact information for a possible follow-up interview, of which 7 actually participated in the interview. The profile of interview participants and findings from the interviews are presented below.

2.1 Interview Participants' Profile

The characteristics of the firms participating in the interviews are summarized in Table 3.6 below.

2.2 Interview Findings

2.2.1 Suppliers Influence on Decision Making for Technology Investments

For the firm's interviewed, suppliers influence their choice in technology when the firm is a producer of technology. The influence is not only linked to the technology supplier use but as one firm stated, "we are careful not to provide all the strategic elements to a same supplier." For firms providing services, suppliers appear to have little impact. It is worth noting that even though these firms considered themselves independent from their suppliers in terms of technology choices, some respondents mentioned that they relied on external and independent consultants to guide their technology choices, as they do not have the critical mass of knowledge internally to make fully informed choices. It should be observed that none of the interview participants operated in a manufacturing industry, which may see an increasing impact of the suppliers on their technology choices.

2.2.2 Clients Influence on Decision Making for Technology Investments

For the firm's interviewed, clients influence their choice in technology. When the firm is a producer of technology, the firm is either technologically ahead of its client and pushes the technology or the firm's products are clearly client driven. For firms providing services, clients have impact to the extent that the firms want to meet their requirements, but not beyond. "We only worry about their requirements." Again, it should be noted that none of the interview participants operated in a manufacturing industry, which may see an increasing impact of the clients on their technology choices.

2.2.3 Competitors Influence on Decision Making for Technology Investments

For the firms interviewed, competitors influence their choice in technology to the extent that firms want to keep up with their competition. In addition, technology-based firms operating in an industry with very few competitors, they "try to collaborate. Whether we are competitors or not, we face the same issues, so there is a lot of information sharing." In the case of firms producing technology and of a small size, competitors can drive them out of a product line, and they are therefore constantly looking for the next market niche. It should be noted that none of the interview participants operated in a manufacturing industry, which may see an increasing impact of the competitors on their technology choices.

2.2.4 Regulators Influence on Decision Making for Technology Investments

For the firms interviewed, regulators influence their choice in technology because of the standards they set for their respective industries, whether in terms of safety or reporting requirements. In the case of technology producers whose products are client driven, regulators influencing the industries of their client also influence where their business is headed. In addition, for international technology-producing firms, if regulators in different countries set different standards, they introduce unfair competition among the international firms, favoring the ones in their domestic market or the ones meeting most closely the requirements of a given market. Similar findings could be expected in the manufacturing industries.

3 Simulation Analysis

The simulation was run three times on each of two configurations, generating a dataset in excess of 70,000 agent observations, which is reduced to 1,536 observations at a lattice point (or subpopulation) level of analysis.[1] Running repeated trials of the same configuration reduces any bias issues and provides a statistically significant sample for analysis. Analysis consists of two stages: graphical analysis and pattern analysis at the lattice point data.

3.1 *Graphical Analysis*

Two simulation variables provide insights into the effects of subpopulation: new venture formation level and probability of new venture formation. New venture formation level provides data on historic performance, while the probability of new venture formation offers insight regarding future system behaviors. To make visual analysis of these outputs easier, the levels of new venture formation and probabilities of new venture formation in Figs. 3.1 and 3.2, respectively, have been tiered using a modified Pareto distribution. The darkest regions represent lattice points where results are 80% or more of the maximum value of the measure on the lattice. The next darkest regions show lattice points where the values are greater than the mean of the measure across lattice points, and the lighter gray indicates lattice points at the 20% (of maximum value) mark. White lattice points represent the lowest performance (<20%). The results presented here are captured at the conclusion of the simulation running for 50 periods.[2]

New venture formation level identifies the number of ventures that are created and survived to reach the incumbent status in a given lattice point. Figure 3.1 shows one of the three runs for each of the two configurations. Other runs within each configuration displayed similar results. The vertical bars separate firms from the three subpopulations:

---

[1] Each configuration generates on the order of 30,600 agents over two runs, which aggregates to 71,200 ±. When these observations are aggregated to measures of regional performance (e.g., number of new venture formation) for this study, the resulting dataset includes 1,536 observations over two configurations (16 × 16 matrix = 256 lattice points × 3 runs/matrix = 768 lattice point observations per configuration × 2 configurations = 1,536 observations).

[2] As mentioned in Annex V, Provance (2010) showed that the simulation became stable at or around 50 periods.

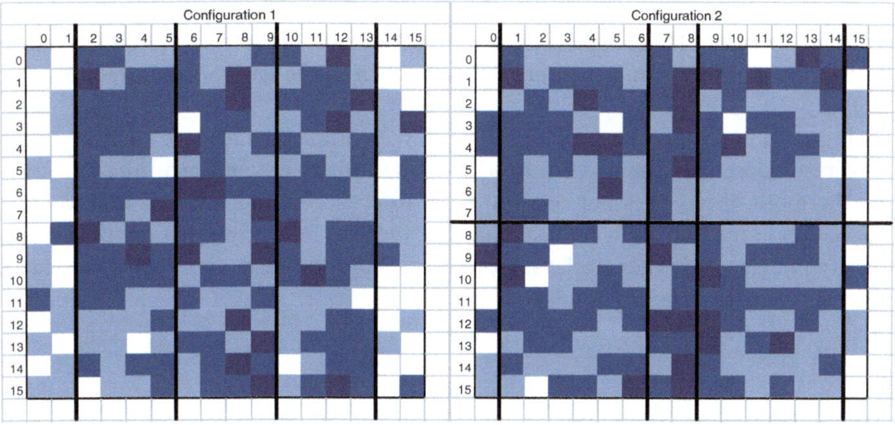

Fig. 3.1 New venture formation levels across configurations and subpopulations

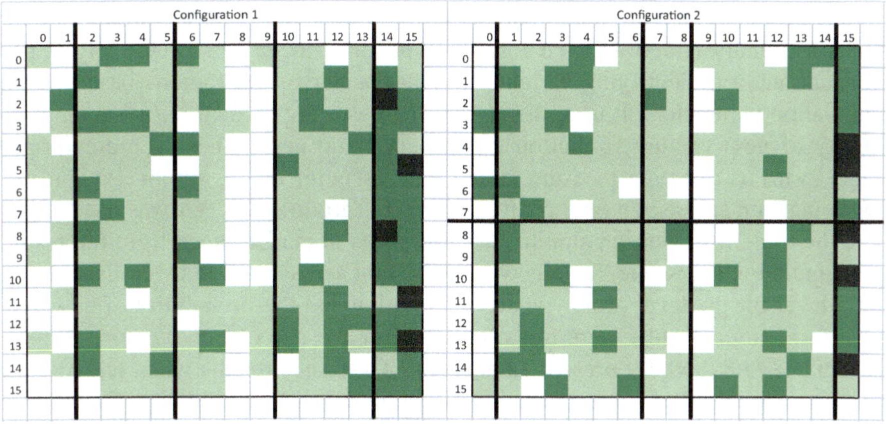

Fig. 3.2 Survival rates for new ventures, expressed as probability of formation occurring across configurations and subpopulations

Configuration 1:	Configuration 2:
Columns 0,1, 14, and 15: TN	Columns 0 and 15: TN
Columns 2–5 and 10–13: TB	Columns 1–6 and 9–14: TB
Columns 6–9: TD	Columns 7–8: TD

From the graphical analysis, it appears that the level of new venture formation is higher in TD than in TB and higher in TB than in TN. This seems to be consistent for both configurations and for both the top and bottom half of the landscapes in configuration 2. However, analyzing the bottom half of configuration 2, in which the early adoption rate has been increased for firms with increased dependence on technology, it does not appear that this factor has an impact on the level of new venture formation. These observations gain additional support from a graphic analysis of the probabilities of new venture formation across the lattice structure presented in Fig. 3.2.

Figure 3.2 depicts the probability of new venture formation by lattice point. The results are consistent with those presented for new venture formation level. Furthermore, there does not seem to be a difference in the level of new venture formation in the two areas representing TN, but Fig. 3.2 shows that the probability of survival seems higher in the region represented on the right of the landscapes. This shows in both configurations.

3.2 *Statistical Analysis*

Since configuration 1 was meant to test the validity of the conceptual model, the statistical analysis focused on configuration 2, which was designed to more closely reflect reality. The statistical analysis corroborated the general findings of the statistical analysis and allowed for more specific conclusions. The analysis that concentrated on identifying variation in patterns across the regions based on the simulation parameters. It used the level of new venture formation (nvsurv), probability of new venture formation (cp_nvsurv), and age of new venture formed (age_form) as outcome measure, along with assignments of an observation to a particular run and region on the lattice structure (group). ANOVA was used to isolate the regions exhibiting significant differences in performance from the rest of the simulation landscape. Results were consistent across runs and significant at the 0.05 level for the level of new venture formation and the probability of new venture formation. Results were not significant for the age of new venture formed. Tables 3.7, 3.8, and 3.9 present an analysis of variance results when testing both the aggregate level of new venture formation (nvsurvive), probability of formation (p_nvsurv), and age of new venture formed (age_form) across regions within each run of configuration 2.

The analysis of the level of new venture formation confirmed differences in the mean of new venture formation levels for in all runs. Post hoc Bonferroni tests demonstrated that TN formation levels were significantly lower than those of TB and TD. They also showed that if TN adopts a lower early adoption rate, its formation levels significantly increase and reach means not significantly different from the ones for TD and TB. The analysis of the probability of new venture formation also exhibited differences in means. Post hoc Bonferroni tests depict similar results to the level for new venture formation. The only difference being that TB appears to have slightly higher (but not significant) survival rates than TD. These statistical results only partially confirm the results of the graphical analysis in that the results do not show significant differences between TD and TB.

Table 3.7 Analysis of variance for levels of new venture formation across regions

Source	SS	df	MS	F	Prob > F
Run 0					
Between groups	**1,117.30738**	**5**	**223.461476**	**5.21**	**0.0001**
Within groups	**10,714.677**	**250**	**42.85870**		
Total	**11,831.9844**	**255**	**46.3999387**		
Bartlett's test for equal variances: $chi^2(5)$ = **7.8622** Prob>Chi^2 = **0.164**					
Run 1					
Between groups	**1,048.1632**	**5**	**209.63264**	**5.12**	**0.0002**
Within groups	**10,227.9579**	**250**	**40.9118316**		
Total	**11,276/1211**	**255**	**44.2200827**		
Bartlett's test for equal variances: $chi^2(5)$ = **4.8666** Prob > Chi^2 = **0.432**					
Run 2					
Between groups	**1,821.37693**	**5**	**364.275387**	**8.25**	**0.0000**
Within groups	**11,044.2324**	**250**	**44.1769298**		
Total	**12,865.6094**	**255**	**50.4533701**		
Bartlett's test for equal variances: $chi^2(5)$ = **7.4136** Prob > chi^2 = **0.192**					

Table 3.8 Analysis of variance for age of new venture formation across regions

Source	SS	df	MS	F	Prob > F
Run 0					
Between groups	**39.71502**	**5**	**7.943004**	**5.00**	**0.0002**
Within groups	**397.204096**	**250**	**1.58881638**		
Total	**436.919116**	**255**	**1.7134083**		
Bartlett's test for equal variances: $chi^2(5)$ = 7.2767 Prob > chi^2 = **0.201**					
Run 1					
Data not available (simulation results returned errors on this variable)					
Run 2					
Between groups	**1.89865864**	**5**	**0.379731727**	**0.23**	**0.9469**
Within groups	**404.283758**	**250**	**1.61713503**		
Total	**406.182417**	**255**	**1.59287222**		
Bartlett's test for equal variances: chi^2 (5) = **7.1987** Prob > chi^2 = **0.206**					

3.3 Summary

This work sought to answer to the following question, which is motivated by the role of real options as a risk management and an uncertainty filtering methodology that helps minimize downside risk and maximize upside potential. The answers to these questions are provided below:

- Why, when, and how are real options used in strategic technology venturing?

Table 3.9 Analysis of variance for probability of new venture formation across regions

Source	SS	df	MS	F	Prob > F
Run 0					
Between groups	**0.97278775**	**5**	**0.19455755**	**12.26**	**0.0000**
Within groups	**3.96781997**	**250**	**0.01587128**		
Total	**4.94060772**	**255**	**0.019374932**		
Bartlett's test for equal variances: chi^2 **(5)** = **7.7993** Prob > chi^2 = **0.168**					
Run 1					
Between groups	**1.17605377**	**5**	**0.235210755**	**16.53**	**0.0000**
Within groups	**3.55768764**	**250**	**0.014230751**		
Total	**4.73374142**	**255**	**0.018563692**		
Bartlett's test for equal variances: chi^2 **(5)** = **2.6524** Prob > chi^2 = **0.753**					
Run 2					
Between groups	**0.95444902**	**5**	**0.190889804**	**10.59**	**0.0000**
Within groups	**4.50606486**	**250**	**0.018024259**		
Total	**5.46051388**	**255**	**0.02141378**		
Bartlett's test for equal variances: chi^2 **(5)** = **6.3503** Prob > chi^2 = **0.274**					

Answer:
This research showed that an environment presenting co-opetitive (simultaneous competition and collaboration) conditions triggers the use of real options (why) that serve to transform the position, posture, and propensity of businesses to innovate, and thus they coevolve (when) into more effective and efficient forms of businesses (co-specialization) (how).

This answer was derived from the answers to the following sub-questions:

- Does the use of real options in decision making for technology investments increase in a co-opetitive environment? (why)

Answer:
Yes, the use of real options increases in a co-opetitive environment when risk and uncertainty increase, in particular the options to abandon an investment or for an investment that create a growth option.

- Does the use of real options in decision making for strategic and nonstrategic technology investments increase with risk and uncertainty? (when)

Answer:
Yes, the use of rescaling and growth options increases for strategic technology investment when risk and uncertainty increase. There is, however, no significant evidence that this applies to nonstrategic technology investments.

- Does the use of real options in decision making increase when the technology investment under consideration leads to new business areas (product, services, and/or processes)? (how)

Answer:
Yes, real-options use increases with risk and uncertainty when the technology investment under consideration leads to new business areas. In particular, options to rescale and the creation of growth options become more important.

3.3.1 Interviews

The interview findings showed that for the set of firms that participated in the interviews, suppliers influence technology choices for technology-producing firms. They also showed that clients impacted technology choices only for companies whose products were clearly client driven and not the results of technology push. Furthermore, competitors were identified as having an impact on technology choices for technology-producing firms; otherwise, firms' main interest was to keep up with what competitors offered. Finally, the findings showed that regulators main impact on technology choices was through standard setting, except for international technology-producing firms whose ability to compete may be affected by regulators adopting different standards in different countries.

3.3.2 Simulation

In addition to providing modeling data to answer these research questions, this work also used an agent-based, knowledge-driven simulation to present evidence supporting the conceptual structure framing this work.

The simulation presented evidence that embracing risk and uncertainty can increase level of new venture formation and probability of new venture formation. However, it also showed that it should be adapted to the risk profile of the firm and that timing is also a factor to be considered. While increasing risk at firm creation increases both the level and probability of new venture formation, increasing the rate of early adoption did not significantly affect it. For more risk-averse firms, lowering the rate of early adoption actually significantly increased the level of new venture formation as well as the probability of new venture formation. Finally, evidence seems to point that there may be an optimal level of risk inclination, as the results from TD and TB were not statistically significant, while the results were statistically significant for TN when compared to either TB or TD (Bonferroni posttest).

4 Validity, Reliability, and Generalizability of the Work

The internal validity of the data will be mostly insured by using negative case sampling, that is, we analyzed with the same level of detail and attention any case that would disconfirm my expectation. The main limitation of internal validity is the lack of representation of TN firms in the survey analysis and interviews.

This research could become more robust by including a representative proportion of TN firms. A research partnership with the NFIB or a similar organization may enhance the representativity of TN firms.

The external validity of the data is less of a concern because the main purpose of the research is explanatory and conclusions drawn from the survey and interviews should be interpreted as such.

Reliability addresses the consistency of measures and/or judgments. When reporting on the interviews, we used low-inference descriptors such as verbatim (Singleton, Royce and Straits 1999). In the case of the interviews, respondent validation (Maxwell 2005) was used: feedback was solicited about data and conclusions from the interview participants. As the quantitative modeling and hypothesis testing was complying with the hypotheses of each of the tests performed, reliability should not represent an issue. The robustness and stability of the simulation were validated by Provance (2010, p. 121).

Because of the exploratory nature of the research, generalizing the simulation findings to the general population can be considered acceptable (Patten 2007), when interpreted within the set parameters, but would otherwise be limited.

Chapter 4
Conclusions

This work was motivated by the role of real options as a risk management and an uncertainty-filtering methodology that helps minimize downside risk and maximize upside potential of a firm's investments. Firms evolve in an uncertain and dynamic environment in which they use the "new knowledge [they] derived through the healthy balance between competition and cooperation involving employees and business partners" (*strategic knowledge co-opetition*, E. G. Carayannis 2009) in the definition of their real options. These real options serve as the basis for their decision making so as to reap the full benefits of the flexibility embedded in their investments. By the exercise of their options, firms have changed the parameters of their previously temporarily stable ecosystem, resulting in a now unstable environment. Having completed the co-opetition process, firms create "new knowledge through a series of interactions and changes at various levels of the organization, spurred by the co-generation and complementary nature of that knowledge," what Carayannis (ibid) coined s*trategic knowledge coevolution*. Through innovation, they also undergo *strategic knowledge co-specialization*, "learning and knowledge which encourages individuals or groups to expand their roles into new areas and new domains, in a complementary and mutually-reinforcing fashion" (ibid). Strategic knowledge co-specialization enables firms to develop *sustainable entrepreneurship* (E. G. Carayannis 2008), that is, "the creation of viable, profitable and scalable firms that engender the formation of self-replicating and mutually enhancing innovation networks and knowledge clusters leading towards what we call robust competitiveness." As such, firms are enabled to develop a temporarily "unfair" competitive advantage as they develop knowledge in new areas and exploit them. In this context, firms now evolve in a temporarily stable environment, s*ustainable entrepreneurship* (ibid), a "state of economic being and becoming that avails systematic and defensible 'unfair advantages' to the entities that are part of the economy and is built on mutually complementary and reinforcing low-, medium and high technology, public and private sector entities (government agencies, private firms, universities, and non-governmental organizations)."

C.M. Sipp and E.G. Carayannis, *Real Options and Strategic Technology Venturing: A New Paradigm in Decision Making*, SpringerBriefs in Business 31,
DOI 10.1007/978-1-4614-5814-2_4, © Springer Science+Business Media New York 2013

To provide empirical evidence in support of this conceptual framework, this work tested with real-world data for the role of real options in decision making involving three types of firms in decreasing order of technology dependence – technology-driven (TD) (where the profit is fully dependent on new technology creation and leveraging), technology-based (TB) (where the profit is enabled and supported by technology), and technology-neutral (TN) (where the profit is almost independent of technology).

This work showed that an environment presenting co-opetitive (simultaneous competition and collaboration) conditions triggers the use of real options (why) that serve to transform the position, posture, and propensity of businesses to innovate and thus they coevolve (when) into more effective and efficient forms of businesses (co-specialization) (how).

1 Value Proposition (So What?)

This book contributes to the literature both in terms of theory and empirical research but also provides evidence that has implications for decision makers faced with the challenge of allocating scarce resources efficiently.

1.1 Contribution to the Theory

This book contributes to the theory by building on previous work on the topic of strategy design and real options, in particular by Bowman and Hurry (1993) who initially framed corporate strategy as a series of real options. This work argues that strategic knowledge serendipity and arbitrage allows companies to better define the content and exercise timing of their real options, as they take part in the co-opetition, coevolution, and co-specialization processes. In doing so, they leverage real options as risk management and uncertainty filters that will allow them to perform better over the longer term.

This book also builds on the research conducted by Carayannis (2008) on strategic knowledge serendipity and arbitrage, co-opetition, coevolution, and co-specialization. It is built on the tenants of SKARSE™ and 3Cs, incorporating real options as an integral part of these processes and extending their application to the formulation and implementation of corporate strategy as a mean of improving firm performance over the longer term.

1.2 Contribution to the Empirical Research

This book also contributes to the empirical research undertaken in the field of real options, little having been conducted on the actual use of real options in decision making (mostly, Bushy and Pitts 1997). This work gathered exploratory data on the use of real options in decision making related to technology investments. It showed that real options gain importance when risk and uncertainty increase, but not for nonstrategic technology investments. It also concluded that creating a growth option matters, followed by the option to rescale a technology investment. The option to abandon only yielded significant results once, while the option to postpone did not yield significant results in any of the proposed cases.

This book also provides a vehicle to test some of Carayannis' (2008) concepts related to co-opetition, coevolution, and co-specialization. It showed that an environment presenting co-opetitive (simultaneous competition and collaboration) conditions triggers the use of real options that serve to transform the position, posture, and propensity of businesses to innovate and thus they coevolve into more effective and efficient forms of businesses (co-specialization).

Finally, it also builds upon and contributes to the newly established strand of research by Provance (2010) based on computer simulation to investigate the role and impact of innovation networks and knowledge clusters on entrepreneurship and innovation. This work conceptualized investments as knowledge-based transactions, which allowed for the incorporation of a real-options lens in Provance's simulation.

1.3 Implications for Policy Makers, Academic Scholars, and Expert Practitioners

The simulation presented evidence that embracing risk and uncertainty can increase level of new venture formation and probability of new venture formation. However, it also showed that it should be adapted to the risk profile of the firm and that timing is also a factor to be considered. While increasing risk at firm creation increases both the level and probability of new venture formation, increasing the rate of early adoption did not significantly affect it. For more risk-averse firms, lowering the rate of early adoption actually significantly increased the level of new venture formation as well as the probability of new venture formation. Finally, evidence seems to point that there may be an optimal level of risk inclination, as the results from the two more risk-embracing groups of firms were not statistically significant.

Real-world data showed that real options gain importance when risk and uncertainty increase for strategic investments. Creating a growth option and having the opportunity to rescale mattered more often than other options. In fact, the option to postpone a technology investment appears not to significantly increase when risk and uncertainty increase.

These conclusions should be taken into account when faced with the challenge of optimizing the allocation of scarce resources with timing, selection, and sequencing decisions. The systematic application of real options could lead to improved cost-benefit and cost-efficiency in the public sector and maximize firm performance and strategic benefits in the private sector.

2 Future Research

Future research opportunities can be grouped into two categories: (1) empirical research that aims at increasing the amount of data collected to be compared and contrasted with existing theories and (2) theory research that can be advanced and refined when informed with empirical and modeling data.

2.1 Increased Wealth of Empirical Data

When designing this work, firms were mostly treated as making decision independently of their location in the value chain of their industry. The simulation moved closer to bridging this gap since it incorporated firms' innovation networks and access to knowledge clusters. The interview data open the path to real-world data collection on this topic but would need more breadth and depth, as well as address manufacturing industries that were not represented in this research. The impact of the supply chain, that is, of suppliers and clients, on decision making and the use of real options could also attract further research, especially in the context of technology venturing.

Although not statistically significant in every case, the simulation results have proven interesting, since the initial result showing that firm performance increases as it takes on more risk is altered when the degree of risk taking is being differentiated between subpopulations at a second decision-making point. The initial relation of performance TD>TB>TN becomes TB>TD>TN. Additional research on this topic could provide a much improved understanding of how much risk is optimal and how to best leverage it depending on the type of strategic decisions the firm has to take.

It would be interesting to test more fully the proposed conceptual model and analyze whether the correlation between the use of real options and superior firm performance over the longer term is empirically validated, both by real-world data and enhanced simulation data.

2.2 Refined Conceptual Framework

The simulation provided data and a testing environment for much deeper analysis. Conclusions drawn are based on initial setup conditions and on a simplified set of modified parameters that allow for analysis. More information could be collected and analyzed from the configurations used in this research in terms of timing and sequencing of decisions, but also many more combinations of parameters could be used to explore various areas of interest that would need refinement, such as exploring the optimal level of risk taking and if this level is conditioned by the risk taking level of a firm at its inception.

3 Impact of the Financial Crisis on the Topic

Real options allow for taking on more risk more successfully. With more traditional approaches to decision making for investments, there is a risk of inappropriate or insufficient valuation and management of risk, resulting in under- or mis-investment. This could be mitigated by deploying real-options methodologies. With increased risk and uncertainty, the more appropriate and rewarding the use of real options can be. A real-options approach allows for a better mitigation and management of risk and therefore increases the return per unit of asset invested. It brings about a higher potential return per unit of asset while minimizing over-, under-, and mis-investing.

Although respondents were asked about how much they valued flexibility after the fact (asked in 2011 about 2008), the results of the survey reported for 2008 can help us understand the impact of the crisis on the current research. In a co-opetitive environment, flexibility is valued today but was not in 2008 (i.e., the results were statistically significant for today, but not for 2008). For strategic technology investment, the investments generating growth options produced statistically significant results, both for today and 2008. For technology investments under a co-specialization (e.g., new business area/product), respondents valued flexibility both today and in 2008. Although the results about 2008 may have been tainted by the a posteriori survey, results seem to point out the following: real options are used for strategic technology investments under coevolution and technology investments under co-specialization (where most expected), but they were not used under co-opetition prior to the crisis, while they are now. With the crisis and corresponding tightening of resources both in the public and private sectors, firms should be interested in managing and leveraging the risks they undertake, so as to maximize the full potential of their investments and more importantly maximize their returns per unit of asset invested.

The systematic use of real options is a possible powerful methodology and device for optimal risk management and risk filter under variable configuration and operationalization of resource allocation. It optimizes the allocation of scarce resources with timing, selection and sequencing decisions, improving cost/benefit and cost/efficiency (public sector view), and maximization of performance and strategic benefit (private sector view).

Although engaging the concepts of real options, this research does not focus on a specific investment valuation methodology, but highlights the relationship between knowledge and risk and rather addresses the management of mindsets, as moving toward a systematic conceptualization of real options represents a different paradigm in decision making.

Glossary

Definitions of terms are included throughout the document, in places where this facilitates the immediate comprehension on the concepts developed in the corresponding sections. Nonetheless, to provide a central repository, the key terms and concepts are also defined below.

Risk Known unknowns (E. G. Carayannis 2000–2009).

Robust competitiveness (E. G. Carayannis 2008) "the creation of viable, profitable and scalable firms that engender the formation of self-replicating and mutually enhancing innovation networks and knowledge clusters."

Strategic knowledge arbitrage (E. G. Carayannis 2008) "the ability to distribute and use specific knowledge for applications other than the intended topic area. More specifically, it refers to the capacity to create, identify, reallocate and recombine knowledge assets more effectively and efficiently to derive, develop and capture non-appropriable, defensible, sustainable and scalable pecuniary benefits" (p. 346).

Strategic knowledge coevolution (E. G. Carayannis 2009) "new knowledge through a series of interactions and changes at various levels of the organization, spurred by the co-generation and complementary nature of that knowledge."

Strategic knowledge co-opetition (E. G. Carayannis 2009) "New knowledge derived through the healthy balance between competition and cooperation involving employees and business partners."

Strategic knowledge co-specialization (E. G. Carayannis 2009) "learning and knowledge which encourages individuals or groups to expand their roles into new areas and new domains, in a complementary and mutually-reinforcing fashion."

Strategic knowledge serendipity (E. G. Carayannis 2008) "the unintended benefits of enabling knowledge to 'spill over' between employees, groups and functional domains ('happy accidents' in learning). More specifically, it describes the capacity to identify, recognize, access and integrate knowledge assets more effectively and efficiently to derive, develop and capture non-appropriable, defensible, sustainable and scalable pecuniary benefits" (p. 346).

C.M. Sipp and E.G. Carayannis, *Real Options and Strategic Technology Venturing: A New Paradigm in Decision Making*, SpringerBriefs in Business 31,
DOI 10.1007/978-1-4614-5814-2, © Springer Science+Business Media New York 2013

Strategic technology investment The technology and/or knowledge derived from the technology investment leads to a substantial change in the business model and value proposition including a shift into a different or emerging sector or business niche. Type of decision making that affects the long-term goals of the business and defines policies (Wysocki 2006). This type of decision making is more likely to rely on disruptive innovation (E. G. Carayannis et al. 2003).

Strategy "The art of devising or employing plans or stratagems toward a goal" or "an adaptation or complex of adaptations (as of behavior, metabolism, or structure) that serves or appears to serve an important function in achieving evolutionary success" (Merriam-Webster 2009). The capacity "to discern the non-obvious faster, better, cheaper and longer" (E. G. Carayannis 2000–2009).

Sustainable entrepreneurship (E. G. Carayannis 2008) A "state of economic being and becoming that avails systematic and defensible 'unfair advantages' to the entities that are part of the economy and is built on mutually complementary and reinforcing low-, medium and high technology, public and private sector entities (government agencies, private firms, universities, and non-governmental organizations)."

Technology In the context of this research, the focus is mainly on "high technology," that is, "scientific technology involving the production or use of advanced or sophisticated devices especially in the fields of electronics and computers" (Merriam-Webster 2009).

Technology-based (TB) Businesses that are technology users and whose profit is enabled and supported by technology. Their business model partly relies on technology use (e.g., e-banking).

Technology-driven (TD) Businesses whose profit is fully dependent on new technology creation and leveraging. Their business model relies on the development and commercialization of technology. For example, this includes but is not limited to businesses defined by North American Industry Classification System (NAICS) codes 334xxx (Computer and Electronic Product Manufacturing) or 517xxx (Telecommunications).

Technology-neutral (TN) Businesses that invest in technology, but their business models do not rely on it. Their profit is almost independent of technology.

Technology venturing Technology investment decision made with the intent of leveraging risk and filtering uncertainty for superior performance.

Uncertainty Unknown unknowns (E. G. Carayannis 2000–2009).

Use of real options in decision making Refers to taking into account the value of flexibility, especially postponement, abandonment, rescaling and growth options (in the sense of choices).

References

Barnett ML (2003) Falling off the fence?: a realistic appraisal of a real options approach to corporate strategy. J Manage Inq 12(2):185

Barnett ML (2005) Paying attention to real options. R & D Manage 35(1):61–72

Bathelt H (2001) The rise of a new cultural products industry cluster in Germany: the case of the Leipzig media industry. Department of Economic and Social Geography Working Paper (06–2001), Johann Wolfgang Goethe-Universität, Frankfurt

Baum JR, Wally S (2003) Strategic decision speed and firm performance. Strateg Manage J 24(11):1107–1129

Benaroch M (2002) Managing information technology investment risk: a real options perspective. J Manage Inf Syst 19(2):43–84

Black F, Scholes M (1973) The pricing of options and corporate liabilities. J Polit Econ 81(3):637

Bowman EH, Hurry D (1987) Strategic options. Working paper 87–20, Reginald Jones Center, the Wharton School, University of Pennsylvania, Philadelphia

Bowman EH, Hurry D (1993) Strategy through the option lens: an integrated view of resource investments and the incremental-choice process. Acad Manage Rev 18(4):760–782

Burt RS (1995) Structural holes: the social structure of competition. Harvard University Press. Cambridge, MA

Bushy JS, Pitts CGC (1997) Real options and capital investment decisions. Manage Account Lond 75:38–39

Cantner U, Hanusch H (2001) Heterogeneity and evolutionary change: empirical conception, findings and unresolved issues. In: Foster J, Metcalfe JS (eds) Frontiers of evolutionary economics: competition, self-organization, and innovation policy. Edward Elgar, Cheltenham, p 228

Carayannis E (1994) The strategic management of technological learning: transnational decision making frameworks and their empirical effectiveness. Published PhD dissertation, School of Management, Rensselaer Polytechnic Institute, Troy

Carayannis E, Stokes R (2000) Strategic management of technological learning. CRC Press, Boca Raton, FL

Carayannis EG (1998) The strategic management of technological learning in project/program management: the role of extranets, intranets and intelligent agents in knowledge generation, diffusion, and leveraging. Technovation 18(11):697–703

Carayannis EG (1999) Fostering synergies between information technology and managerial and organizational cognition: the role of knowledge management. Technovation 19(4):219–231

Carayannis EG (2000–2009) GWU SB Lectures, The George Washington University. Washington, DC

C.M. Sipp and E.G. Carayannis, *Real Options and Strategic Technology Venturing: A New Paradigm in Decision Making*, SpringerBriefs in Business 31,
DOI 10.1007/978-1-4614-5814-2,

Carayannis EG (2004) GWU MSTI Doctoral Seminar, The George Washington University. Washington, DC
Carayannis EG (2008) Knowledge-driven creative destruction, or leveraging knowledge for competitive advantage Strategic knowledge arbitrage and serendipity as real options drivers triggered by co-opetition, co-evolution and co-specialization. Ind High Educ 22(6):343–353
Carayannis EG (2009) Firm evolution dynamics: towards sustainable entrepreneurship and robust competitiveness in the knowledge economy and society. Int J Innov Reg Dev 1(3):235–254
Carayannis EG, Assimakopoulos D, Kondo M (2008) Innovation networks and knowledge clusters. Palgrave McMillan, New York, NY
Carayannis EG, Campbell DFJ (2006) Knowledge creation, diffusion, and use in innovation networks and knowledge clusters. Praeger, Westport, CT
Carayannis EG, Campbell DFJ (2009) 'Mode 3'and'Quadruple Helix': toward a 21st century fractal innovation ecosystem. Int J Technol Manage 46(3):201–234
Carayannis EG, Gonzalez E (2003) Creativity and innovation=competitiveness? When, how and why. In: International handbook on innovation. Elsevier Science, Boston, pp 587–605
Carayannis EG, Gonzalez E, Wetter J (2003) The nature and dynamics of discontinuous and disruptive innovations from a learning and knowledge management perspective. In: The international handbook on innovation. pp 115–138 Larisa Shavinina (Ed). Elsevier Science Ltd. Oxford, UK
Carayannis EG, Kaloudis A, Mariussen Å (2008b) Diversity in the knowledge economy and society: heterogeneity, innovation and entrepreneurship. Edward Elgar, Cheltenham
Carayannis EG, Sipp C (2010) Why, when, and how are real options used in strategic technology venturing? J Knowl Econ: Springer 1(2):70–85
Christensen CM (1997) The innovator's dilemma: when new technologies cause great firms to fail. Harvard Business School Press, Boston
Christensen CM, Johnson M, Dann J (2002) Disrupt and prosper. Optimizemag.com, Nov: 41–48
Cimoli M, Katz J (2003) Structural reforms, technological gaps and economic development: a Latin American perspective. Ind Corp Change 12(2):387
Cohen MD, March JG, Olsen JP (1972) A garbage can model of organizational choice. Adm Sci Q 17(1):1–25
Cohen WM, Levinthal DA (1990) Absorptive capacity: a new perspective on learning and innovation. Adm Sci Q 35(1)
Copeland TE, Antikarov V (2001) Real options: a practitioner's guide. WW Norton, Oakland, CA
Covin JG, Slevin DP (2002) The entrepreneurial imperatives of strategic leadership. In: Strategic entrepreneurship: creating a new mindset. pp 309–327. Michael A. Hitt, R. Duane Ireland, S. Michael Camp and Donald L. Sexton (Eds). Blackwell Publishers. Malden, MA
Cox JC, Ross SA, Rubinstein M (1979) Option pricing: a simple approach. J Financ Econ 7:229–263
Creswell JW (2003) Research design: qualitative, quantitative, and mixed methods approaches, 2nd edn. Sage, Thousand Oaks
Das TK, Teng BS (1999) Cognitive biases and strategic decision processes: an integrative perspective. J Manage Stud 36(6):757–778
Davis B (2000) The power of knowledge pattern recognition. Kaieteur Institute for Knowledge Management, Toronto
Day GS, Schoemaker PJH (2000) Avoiding the pitfalls of emerging technologies. Calif Manage Rev 42(2):8–33
Dixit A (1992) Investment and hysteresis. J Econ Perspect Vol 6(1):107–132
Dixit AK, Pindyck RS, Davis GA (1994) Investment under uncertainty. Princeton University Press, Princeton
Dosi G, Teece D, Winter S (1992) Toward a theory of corporate coherence: preliminary remarks. Technology and Enterprise in a Historical Perspective: theory and history of technology and business firms: the microeconomics of industrial development/Giovanni Dosi, Renato Giannetti, and Pier Angelo Toninelli The concept of the industrial revolution/Sidney Pollard Science and technology in the twentieth century/Nathan Rosenberg The dynamics of techno-

logical change: salients, critical problems, and industrial revolutions/Thomas P. Hughes Business organization and competitive advantage: capitalist transformations in the twentieth century/William Lazonick The roles of firms in technical advance: a perspective from evolutionary theory/Richard Nelson Toward a theory of corporate coherence: preliminary remarks/ Giovanni Dosi, David J. Teece, and Sidney Winter Some foundations for a theory of the large innovating firm/Keith Pavitt, 185

Drucker PF (1968) The age of discontinuity: guidelines to our changing economy. Harper & Row, New York

Durand R, Gomez PY, Monin P (2001) Real options strategies, European Entrepreneurial Learning. EM Lyon. Lyon, France

Eisenhardt KM (1989) Making fast strategic decisions in high-velocity environments. Acad Manage J Vol 32(3):543–576

Episcopos A (1995) Evidence on the relationship between uncertainty and irreversible investment. Q Rev Econ Finance 35(1):41–52

Freeman C (1985) The economics of innovation. IEE Proc Phys Sci, Meas Instrum, Manage Educ, Rev 132(4):213–221

Fuchs M (2001) Von der“ lernenden Region” zur“ lernenden Organisation”. INEF

Grenadier SR, Weiss AM (1997) Investment in technological innovations: an option pricing approach. J Financ Econ 44(3):397–416

Grewal R, Tansuhaj P (2001) Building organizational capabilities for managing economic crisis: the role of market orientation and strategic flexibility. J Mark 65(2):67–80

Hull J (2008) Options, futures, and other derivatives. Pearson Prentice Hall, India

Hurry D (1993) Restructuring in the global economy: the consequences of strategic linkages between Japanese and US firms. Strateg Manage J 14:69–82

Hurry D, Miller AT, Bowman EH (1992) Calls on high-technology: Japanese exploration of venture capital investments in the United States. Strateg Manage J 13(2):85–101

Hymer SH (1960) The international operations of national firms: a study of foreign direct investment. PhD dissertation, Massachusetts Institute of Technology

Ireland RD, Hitt MA, Sirmon DG (2003) A model of strategic entrepreneurship: the construct and its dimensions. J Manage 29(6):963

Jensen MC, Ruback RS, Field S, Park TC (1983) The market for corporate control: the scientific evidence. J Financ Econ 11:5–50

Kao J (1996) Jamming: the art and discipline of business creativity. HarperCollins, New York

Kedia BL, Nordtvedt R, Perez LM (2002) International business strategies, decision-making theories, and leadership styles: an integrated framework. Compet Rev: Int Bus J Inc J Glob Compet 12(1):38–52

Keen PGW, Morton MSS (1978) Decision support systems: an organizational perspective. Addison Wesley, Reading

Klein GA (1999) Sources of power: how people make decisions. The MIT Press, Cambridge

Kline SJ, Rosenberg N (1986) An overview of innovation. In: The positive sum strategy: Harnessing technology for economic growth. pp 275–305. Ralph Landau and Nathan Rosenberg (Eds). National Academy Press, Washington, DC

Koestler A (1964) Bisociation in creation. In: Rothenberg A, Hausman E (eds) The creativity question. Duke University Press, Durham

Kogut B (1991) Joint ventures and the option to expand and acquire. Manage Sci 37(1):19–33

Kogut B, Kulatilaka N (1994) Operating flexibility, global manufacturing, and the option value of a multinational network. Manage Sci Vol 40(1):123–139

Kogut B, Kulatilaka N (2001) Capabilities as real options. Organ Sci 12(6):744–758

Krabuanrat K, Phelps R (1998) Heuristics and rationality in strategic decision making: an exploratory study. J Bus Res 41(1):83–93

Lawson C, Lorenz E (1999) Collective learning, tacit knowledge and regional innovative capacity. Reg Stud 33(4):305–317

Lincoln YS, Guba EG (1985) Naturalistic inquiry. Sage, Beverly Hills

Lindblom CE (1959) The science of “ muddling through”. Public Adm Rev 19(2):79–88

Loasby BJ (1998) The organisation of capabilities. J Econ Behav Organ 35(2):139–160
Loughran T, Vijh AM (1997) Do long-term shareholders benefit from corporate acquisitions? J Financ 52(5):1765–1790
Luehrman TA (1998) Strategy as a portfolio of options. Harv Bus Rev 76(5):89–99
Lundvall BA (1988) Innovation as an interactive process: from user-producer interaction to the national system of innovation. In: Technical change and economic theory. pp 349–369, Giovanni Dosi, Christopher Freeman, Richard Nelson and Luc Soete (Eds). Pinter Publishers, London, UK
March JG (1978) Bounded rationality, ambiguity, and the engineering of choice. Bell J Econ 9(2):587–608
March JG (1991) Exploration and exploitation in organizational learning. Organ Sci 2(1):71–87
March JG, Simon HA, Guetzkow H (1958) Organizations, Wiley, USA, NJ
Maxwell JA (2005) Qualitative research design: an interactive approach. Sage, Thousand Oaks
Mayer P (2007) L'organisation détraquée. Revue française de gestion (2007/4), 69–83
McGrath RG (1997) A real options logic for initiating technology positioning investments. Acad Manage Rev 22(4):974–996
McGrath RG (1999) Falling forward: real options reasoning and entrepreneurial failure. Acad Manage Rev 24(1):13–30
McGrath RG, Boisot M (2003) Real options reasoning and the dynamic organization: strategic insights from the biological analogy. In: Leading and managing people in the dynamic organization. p 201. Randall S. Peterson and Elizabeth A. Mannix (Eds). Lawrence Erlbaum Associates Publishers. Mahwah, NJ
McGrath RG, Ferrier WJ, Mendelow A (2004) Real options as engines of choice and heterogeneity. Acad Manage Rev 29(1):86–101
McKinseyQuarterly (1998) Industrial venture capitalism: sharing ownership to create value. McKinsey Q (1): 26–34
Merriam-Webster (ed) (2009) Merriam-Webster Online http://www.merriam-webster.com/
Merton RC (1973) Theory of rational option pricing. Bell J Econ Manag Sci 141–183
Milliken FJ (1987) Three types of perceived uncertainty about the environment: state, effect, and response uncertainty. Acad Manage Rev 12(1):133–143
Mitroff II, Linstone HA (1993) The unbounded mind. Oxford University Press, Oxford
Moran P, Ghoshal S (1999) Markets, firms, and the process of economic development. Acad Manag Rev Vol 24(3):390–412
Myers SC (1977) Determinants of corporate borrowing* 1. J Financ Econ 5(2):147–175
Nonaka I, Takeuchi H (1995) The knowledge-creating company: how Japanese companies create the dynamics of innovation. Oxford University Press, New York
Nunnally JC, Bernstein IH (1978) Psychometric theory. McGraw-Hill, New York
O'Brien JP, Folta TB, Johnson DR (2003) A real options perspective on entrepreneurial entry in the face of uncertainty. Manage Decis Econ Vol 24(8):515–533
Owen-Smith J, Powell WW (2002) Knowledge networks in the Boston biotechnology community. Conference on 'Science as an Institution and the Institutions of Science' University of Siena. Siena, Italy
Patten ML (2007) Understanding research methods: an overview of the essentials. Pyrczak, Los Angeles, CA
Pedhazur EJ, Schmelkin LP (1991) Measurement, design, and analysis: an integrated approach. Lawrence Erlbaum.
Penrose E (1959) The theory of the growth of the firm. Sharpe, New York
Penrose ET (1995) The theory of the growth of the firm (1959). Oxford University Press, Oxford
Pindyck RS (1991) Irreversibility, uncertainty, and investment. J Econ Lit 29(3):1110–1148
Porter M (1990) Competitive advantage of Nations. Harvard Business Review, Cambridge, MA
Provance M (2010) Tying it together: three essays on the roles of knowledge, entrepreneurial action, and institutions in the formation of new ventures. George Washington University, Washington, DC

Romer PM (1986) Increasing returns and long-run growth. J Polit Econ 94(5):1002
Rosenberg N (1982) Inside the black box: technology and economics. Cambridge University Press, New York
Sanchez R (1993) Strategic flexibility, firm organization, and managerial work in dynamic markets: a strategic options perspective. Adv Strateg Manage 9(1993):251–291
Santos JRA (1999) Cronbach's alpha: a tool for assessing the reliability of scales. J Ext 37(2):1–5
Schumpeter JA (1942) Capitalism, socialism and democracy. Harper & Row, New York
Schumpeter JA (1934) The theory of economic development. Harvard University Press, Cambridge (originally published in German in 1911)
Schwartz ES, Trigeorgis L (2001) Real options and investment under uncertainty: an overview. In: Real options and investment under uncertainty. pp 1–16, Eduardo S. Schwartz and Lenos Trigeorgis (Eds). MIT Press. Cambridge, MA
Semadeni M (2001) Toward a theory of knowledge arbitrage. In: Current trends in management consulting (PB). p 43, Anthony F. Buono (Ed). Information Age Publishing Inc. Charlotte, NC
Simon HA (1955) A behavioral model of rational choice. Q J Econ 69(1):99–118
Simon HA (1979) Rational decision making in business organizations. Am Econ Rev Vol 69 (4):493–513
Simon HA (1987) Making management decisions: the role of intuition and emotion. Acad Manage Exec 1(1):57–64
Simon HA (1991) Bounded rationality and organizational learning. Organ Sci 2(1):125–134
Singleton, Royce A, Jr., Straits BC (1999) Approaches to social research. 3d ed. New York: Oxford University Press
Smit H, Trigeorgis L (2006) Strategic planning: valuing and managing portfolios of real options. R&D Manage 36(4):403–419
Smith A (1776) An inquiry into the nature and causes of the wealth of nations. Methuen, London
Smith JE, Nau RF (1995) Valuing risky projects: option pricing theory and decision analysis. Manage Sci 41(5):795–816
Smith KW, Triantis AJ (1995) The value of options in strategic acquisitions. Real Options in Capital Investment; Models, Strategies, and Applications, Praeger, Westport Connecticut.
Smith KG, Di Gregorio D (2002) Bisociation, discovery, and the role of entrepreneurial action. In: Strategic entrepreneurship: creating a new mindset. pp 129–150. Michael A. Hitt, R. Duane Ireland, S. Michael Camp and Donald L. Sexton (Eds). Blackwell Publishers. Malden, MA
Spender JC (1996) Making knowledge the basis of a dynamic theory of the firm. Strateg Manage J 17:45–62
Storper M, Venables AJ (2004) Buzz: the economic force of the city. J Econ Geogr 4(4):351–370
Tong T, Reuer J, Building MC (2004) Corporate investment decisions and the value of growth options. Paper presented at the Proceedings of the annual international conference on real options, Montreal, Canada
Trautwein F (1990) Merger motives and merger prescriptions. Strateg Manage J 11(4):283–295
Tushman ML, O'reilly CA (1996) Ambidextrous organizations: managing evolutionary and revolutionary change. Calif Manage Rev Vol 38(4):8–30
Uzzi B (1996) The sources and consequences of embeddedness for the economic performance of organizations: the network effect. Am Sociol Rev 61(4):674–698
Uzzi B (1997) Social structure and competition in interfirm networks: the paradox of embeddedness. Adm Sci Q 42(1)
Wysocki RK (2006) Effective software project management. Wiley, New York
Young AA (1928) Increasing returns and economic progress. Econ J 38(152):527–542

Annex I
Bibliography on Real Options and Strategic Decision Making

Abbott A, Banerji K (2003) Strategic flexibility and firm performance: the case of US based transnational corporations. Glob J Flex Syst Manage 4(1):1–8

Acedo FJ, Barroso C, Galan JL (2006) The resource-based theory: dissemination and main trends. Strateg Manage J 27(7):621–636

Adner R, Levinthal DA (2004) Real options and real tradeoffs. Acad Manage Rev 29(1):120–126

Albors-Garrigos J, Hervas-Oliver JL, Marquez PB (2008) When technology innovation is not enough, new competitive paradigms, revisiting the Spanish ceramic tile sector. Int J Technol Manage 44(3):406–426

Alessandri TM, Lander DM, Bettis RA (2007) Strategic implications of valuation methods. Adv Strateg Manage 24:473–499

Alsos GA, Kolvereid L (1998) The business gestation process of novice, serial, and parallel business founders. Entrep: Theory Pract 22(4):101–114

Amado J, Montes P, Europeo M. Recursos de Tecnología de la Informaci6n y Desempeño Organizativo: El Rol Mediador de la Capacidad de Agilidad Empresarial

Anand J, Oriani R, Vassolo RS (2007) Managing a portfolio of real options. Adv Strateg Manage 24:275–303

Andersen TJ (2004) Integrating decentralized strategy making and strategic planning processes in dynamic environments. J Manage Stud 41(8):1271–1299

Andersen TJ (2005) Capital structure, environmental dynamism, innovation strategy, and strategic risk management. Center for Strategic Management and Globalization, Copenhagen, Denmark

Andersen TJ (2005) The performance and risk management implications of multinationality: an industry perspective. Center for Strategic Management and Globalization, Copenhagen, Denmark

Andersen TJ (2006) Catastrophe risk analysis and disaster financing: a country perspective38. In Perspectives on Strategic Risk Management, Torben Juul Andersen (Ed). Copenhagen Business School Press. Copenhangen, Denmark

Andersen TJ (2008) Multinational performance and risk management effects: capital structure contingencies. Center for Strategic Management and Globalization, Copenhagen, Denmark

Anderson MH (2006) How can we know what we think until we see what we said?: a citation and citation context analysis of Karl Weick's the social psychology of organizing. Organ Stud (01708406), 27(11): 1675–1692

Anderson TJ (2000) Real options analysis in strategic decision making: an applied approach in a dual options framework. J Appl Manage Stud 9(2): 235–255

Ansari VA (2006) Transaction costs, resource-based and real options views on vertical integration an appraisal. The Journal of Indian Management & Strategy 8M, 11(4):30–34

C.M. Sipp and E.G. Carayannis, *Real Options and Strategic Technology Venturing: A New Paradigm in Decision Making*, SpringerBriefs in Business 31,
DOI 10.1007/978-1-4614-5814-2,

Arend RJ (2004) Volatility-based effects on shareholder value: alliance activity in the computing industry. J Manage 30(4):487–508

Arend RJ (2006) SME-supplier alliance activity in manufacturing: contingent benefits and perceptions. Strateg Manage J 27(8):741–763

Arend RJ, Amit R (2005) Selection in strategic alliance activity: effects on firm performance in the computing industry. Eur Manage J 23(4):361–381

Arend RJ, Mantecon T (2008) Market reactions to asset sales: effects of the Joint Venture Prerequisite. Eur Manage J 26(2):105–121

Arend RJ, Seale DA (2005) Modeling alliance activity: an iterated prisoners' dilemma with exit option. Strateg Manage J 26(11):1057–1074

Arikan A, Arikan I, Kiousis P (2002) Technological investment decisions: implications for real options logic, Paper presented at the Academy of Management. Denver, CO

Armstrong CE, Drnevich PL, Newbert SL (2009) Conversations in competitive advantage: a bibliographic analysis of the major research streams and their influence on the field, Atlanta Competitive Advantage Conference Paper. Atlanta, GA

Armstrong CP, Sambamurthy V (1999) Information technology assimilation in firms: the influence of senior leadership and IT infrastructures. Inf Syst Res 10(4):304–327

Arrègle JL (2006) Analyse "Resource Based" et identification des actifs stratégiques. Revue française de gestion 160(2006/1):241–259

Arya B, Lin Z (2007) Innovation sourcing decisions of hi-tech firms: an embeddedness perspective. Strateg Manage Rev 1(1):1

Ashill NJ, Jobber D (2009) Measuring state, effect, and response uncertainty: theoretical construct development and empirical validation. J Manage 36(5):1278–1308

Atkins M, Anderson A (1999) Risk, uncertainty and the entrepreneurial vision. Risk Manage Vol 1(2):35–48

Avadikyan A, Llerena P (2009) The hybrid vehicle as a transitional investment strategy towards a more sustainable transport system: a real option reasoning, EAEPE Conference. Amsterdam, The Netherlands

Badders B, Clark LC, Wright PM (2007) Uncertainty and human capital decisions: traditional valuation methods and real options logic. CAHRS Working Paper Series 460

Baden-Fuller C, Dean A, McNamara P, Hilliard B (2006) Raising the returns to venture finance. J Bus Ventur 21(3):265–285

Bailey J, Ford C (1996) Management as science versus management as practice in postgraduate business education. Bus Strateg Rev 7(4):7–12

Banik M, Westgren RE (2004) A wealth of failures: sensemaking in a pharmaceutical R&D pipeline. Int J Technol Intell Plan 1(1):25–38

Barnett ML (2003) Falling off the Fence?: a realistic appraisal of a real options approach to corporate strategy. J Manage Inq 12(2):185–196

Barnett ML (2005) Paying attention to real options. R&D Manage 35(1):61–72

Barnett ML (2007) Stakeholder influence capacity and the variability of financial returns to corporate social responsibility. Acad Manage Rev (AMR) 32(3):794–816

Barnett ML (2008) An attention-based view of real options reasoning. Acad Manage Rev (AMR) 33(3):606–628

Barney JB, Lee W (1998) Governance under uncertainty: transaction costs, real options, learning, and property rights. Paper presented at the annual meeting of the academy of management, San Diego

Bausch A (2003) Unternehmungszusammenschlüsse.: Strategien und Strukturen für kooperatives und akquisitorisches Wachstum in Industrieunternehmungen: Duv

Becker M (2005) Controlling von Internationalisierungsprozessen. Gabler Verlag

Benaroch M (2002) Managing information technology investment risk: a real options perspective. J Manage Inf Syst 19(2):43–84

Berndt E (2005) Toyota: Zur Arbeit an der automobilen Zukunft seit den 1990er Jahren. Leipziger Universitätsverlag

Berndt E, Metzner A (2004) Nichts Neues oder Wie groß ist die Varianz in der Automobilindustrie? Thesen zur Entwicklung alternativer Antriebe in Automobilen
Berry, H (2010) Why Do Firms Divest?. Organization Science 21(2):380–396
Bethel JE, Liebeskind JP (1998) Diversification and the legal organization of the firm. Organ Sci 9(1):49–67
Bettis RA, Hitt MA (1995) The new competitive landscape. Strateg Manage J 16(S1):7–19
Bhattacharya M, Wright P (2000) Recognizing risk in human capital investments: a real options approach to strategic human resource management. CAHRS Working Paper Series, 98
Bhattacharya M, Wright P (2005) Options for human capital acquisition. Int J Hum Resour Manage 16(6):929–948
Boccardelli P, Magnusson MG, Oriani R (2008) Distinctiveness and flexibility of resources: a study of the motion picture industry. In Strategic Management Society conference, Cologne, Germany
Bogner WC, Barr PS (2000) Making sense in hypercompetitive environments: a cognitive explanation for the persistence of high velocity competition. Organ Sci 11(2):212–226
Boudier F (2007) Production mondialisée: du besoin d'un nouveau triptyque" concept, mesure, théorie. La Mesure de la Mondialisation, Cahiers du GEMDEV 31:99–114
Bowman EH, Moskowitz GT (2001) Real options analysis and strategic decision making. Organ Sci 12(6):772–777
Bowman EH, Singh H, Thomas H (2002) The domain of strategic management: history and evolution. In Handbook of Strategy and Management. M Pettigrew, Richard Whittington, Howard Thomas (Eds). Sage Publications. Thousand Oaks, CA. 31–51
Brau JC, Sutton NK, Hatch NW (2008) Dual-track versus single-track sell-outs: an empirical analysis of competing harvest strategies. J Bus Ventur 25(4):389–402
Bräutigam J, Esche C, Mehler-Bicher A. Uncertainty as a key value driver of real options. Paper presented at the fifth conference on real options: theory meets practice
Brits J, Botha GHK (2006) Conceptual framework for modeling business capabilities, Doctoral dissertation, Tshwane University of Technology, Pretoria, South Africa
Brouthers KD, Brouthers LE, Werner S (2008) Real options, international entry mode choice and performance. J Manage Stud 45(5):936–960
Bruton GD, Dess GG, Janney JJ (2007) Knowledge management in technology-focused firms in emerging economies: caveats on capabilities, networks, and real options. Asia Pac J Manage 24(2):115–130
Buckley A, Tse K (1996) Real operating options and foreign direct investment: a synthetic approach. Eur Manage J 14(3):304–314
Bukszar E (1999) Strategic bias: the impact of cognitive biases on strategy. Can J Adm Sci/Revue Canadienne des Sciences de l'Administration 16(2):105–117
Burger-Helmchen T (2004) How do real options come into existence? a step toward an option-based theory of the firm, Bureau d'Economie Theorique et Appliquee, University Louis Pasteur, Strasbourg, France
Burger-Helmchen T (2007) Justifying the origin of real options and their difficult evaluation in strategic management, Schmalenbach Business Review, 59
Burger-Helmchen T (2007) Les dangers d'une approche financière des options réelles. Revue française de gestion 170(1):59–74
Calabrese A, Mancuso P (2002) Real options method (ROM): a cognitive model for supporting strategy creation processes in uncertain competitive environments. Braz Electron J Econ 5(1)
Calabrese A, Gastaldi M, Ghiron NL (2005) Real option's model to evaluate infrastructure flexibility: an application to photovoltaic technology. Int J Technol Manage 29(1):173–191
Calhoun MA (2006) Seeking meaningful country-risk information: a wealth of information, a void in understanding. In Perspectives on Strategic Risk Management, Torben Juul Andersen (Ed). Copenhagen Business School Press. Copenhangen, Denmark
Caloghirou Y, Ioannides S, Vonortas N, Greece A (2003) Research joint ventures. J Econ Surv 17(4):541–570
Canals J (2000) Managing corporate growth. Oxford University Press, Oxford

Carlsson C, Fullér R, Heikkilä M, Majlender P (2007) A fuzzy approach to R&D project portfolio selection. Int J Approx Reason 44(2):93–105

Casselman RM (2004) What you don't know can hurt you: towards an integrated theory of knowledge and corporate risk, Department of Management Working Paper. The University of Melbourne. Melbourne, Australia

Cattani G (2006) Technological pre-adaptation, speciation, and emergence of new technologies: how Corning invented and developed fiber optics. Ind Corp Change 15(2):285–318

Chamanski A, Waagø SJ (2003) Critical success factors of new, technology-based firms. In small firms and economic development in developed and transition economies: a reader. David Anthony Kirby, Anna Watson (Eds). 45, Ashgate Publishing Company, Burlington, VT

Chambers DW, Bergstrom R (2004) Financial management and dental school equity, Part II: tactics. J Dent Educ 68(4):439–445

Chang SJ (1995) International expansion strategy of Japanese firms: capability building through sequential entry. Acad Manage J 38:383–407

Chen S (2003) Valuing intellectual capital using game theory. J Intell Cap 4:191–201

Chi T (2000) Option to acquire or divest a joint venture. Strateg Manage J 21:665–687

Chien, Shih-ping, Wu, Wann-Yih (2008) The antecedents and consequences of dynamic resource management model for organizational competitiveness. National Cheng Kung University (NCKU), Department of Business Administration & Institute of International Business, School of Evening and Graduate Studies, Taiwan

Choi YR, Dean Shepherd A (2002) Honeymoons and the entrepreneurial process: a real options perspective. Research paper, School of Business, Singapore Management University and College of Business and Administration, University of Colorado at Boulder

Choi YR, Shanley MT (2000) A new venture's honeymoon period: knowledge, resources, and real options reasoning. Proc Acad Manage

Clemente R, Caulliraux H, Meirelles L, Proença A (2008) A reference framework to support absorptive capacity development 17th - International Conference on Management of Technology - IAMOT; Dubai, U.A.E

Clemons EK, Gu B (2003) Justifying contingent information technology investments: balancing the need for speed of action with certainty before action. J Manage Inf Syst 20(2):11–48

Coelho F, Easingwood C (2003) Multiple channel structures in financial services: a framework. J Financ Serv Mark 8(1):22–34

Coelho F, Easingwood C (2005) Determinants of multiple channel choice in financial services: an environmental uncertainty model. J Serv Mark 19(4):199–211

Coff RW, Laverty KJ (2001) Dilemmas when exercising real options on knowledge-based assets: escalation of commitment vs. killing the golden, Emory University, Atlanta, GA

Coff RW, Laverty KJ (2007) Real options meet organizational theory: coping with path dependencies, agency costs, and organizational form. Adv Strateg Manage 24:333–361

Collan M, Shuhua L (2003) Fuzzy logic and intelligent agents: towards the next step of capital budgeting decision support. Ind Manage Data Syst 103(6):410

Cote D (1999) Power strategy formation in a human service organization undergoing turbulent change, Doctoral Dissertation. University of Montreal, Montreal, Canada

Cui X (2007) RFID adoption timing: the application of option pricing model, Chinese University of Hong Kong

Cunill OM, Forteza CM, Miralles MR (2008) Valuing growth strategy management by hotel chains based on the real options approach. Tour Econ 14(3):511–526

Cuypers IRP, Martin X (2006) What makes and what does not make a real option? A study of international joint ventures. (2006) Academy of Management Best Paper Proceedings

Cuypers IRP, Martin X (2007) Joint ventures and real options: an integrated perspective. Adv Strateg Manage 24:103–144

Daignault M, Shepherd M, Marche S, Watters C (2002) Enabling trust online. Paper presented at the Proceedings of the third international symposium on electronic commerce (ISEC'02), Washington, DC

Dardour I. Docteur de l'Ecole des Mines de Paris.
Das TK, Bing-Sheng T (1998) Between trust and control: developing confidence in partner cooperation in alliances. Acad Manage Rev 23(3):491–512
Day GS, Schoemaker PJH (2000) Avoiding the pitfalls of emerging technologies. Calif Manage Rev 42(2):8–33
de Andrés Alonso P, de la Fuente Herrero G (1998) Modelos Financieros de Valoración de Estrategias Empresariales: Evolución y Perspectivas. In Anales de estudios económicos y empresariales (No. 13, pp. 129–146)
de Andrés Alonso P, Azofra-Palenzuela V, De la Fuente-Herrero G, Esgueva AV (2006) The real options component of firm market value: the case of the technological corporation. J Bus Financ Account 33(1–2):203–219
De Andrés-Alonso P, Azofra-Palenzuela V, De La Fuente-Herrero G (2005) Real options as a component of the market value of stocks: evidence from the Spanish Stock Market. Appl Econ 37(14):1673–1691
de Doctorado T. Las opciones reales en la decisión de inversión. Propuesta y aplicación de un modelo de valoración al caso de una multinacional española
de Meyer A (2008) Technology strategy and China's technology capacity building. J Technol Manage China 3:17
Dean A, Baden-Fuller C (2003) Market entry, pricing decisions and options contracts, Seventh Annual Real Options Conference, Washington, DC
Dempster A (2006) Valuing announcement options. Eur Manage J 24(2–3):117–127
Denison CA (2009) Real options and escalation of commitment: a behavioral analysis of capital investment decisions. Account Rev 84(1):133–155
Denning KC, Hulburt H, Ferris SP (2006) Risk and wealth effects of US firm joint venture activity. Rev Financ Econ 15(3):271–285
Dess GG, Ireland RD, Zahra SA, Floyd SW, Janney JJ, Lane PJ (2003) Emerging issues in corporate entrepreneurship. J Manage 29(3):351
Devers CE, Wiseman RM, Holmes RM (2007) The effects of endowment and loss aversion in managerial stock option valuation, Academy of Management Journal, 50(1):191–208
Dhanaraj C (2006) Research methodology in strategy and management/research methodology in strategy and management. Acad Manage Rev 31(2):497–500
Di Zhao RZ (2008) The application of the thought of strategic options in enterprise strategic change. In Business and Information Management, 2008. ISBIM'08. International Seminar on (Vol. 2, pp. 345–348). IEEE
Dibrell C, Davis PS, Craig J (2008) Fueling innovation through information technology in SMEs. J Small Bus Manage 46(2):203–218
Doh JP, Pearce JA (2004) Corporate entrepreneurship and real options in transitional policy environments: theory development. J Manage Stud 41(4):645–664
Durand R, Gomez PY, Monin P (2002) Théorie des options et management stratégique, Revue Française de Gestion n° 137 pp. 25–40
Durand R, Gomez PY, Monin P (2006) Le management stratégique face à la théorie des options. Revue française de gestion 160(2006/1):159–176
Easingwood C, Coelho F (2003) Single versus multiple channel strategies: typologies and drivers. Serv Ind J 23(2):31–46
Edelmann J, Sintonen S (2006) Adoption of electronic invoicing in Finnish SMEs: two complementary perspectives. Int J Enterp Netw Manage 1(1):79–98
Edelmann J, Laaksonen P, Sandstrom J (2004) Facing the future: competitive situation in telecommunications in terms of real options. Management of Technology: Internet Economy: Opportunities and Challenges for Developed and Developing Regions of the World, 69
Edelmann J, Bergman J, Jantunen A (2005) Managing R&D by normative scenarios. Int J Foresight Innov Policy 2(1):69–83
Ehrmann T, Cochet O. Is entrepreneurship a flexible strategy process? Downside risk implications of EM and EO

El Sawy OA, Malhotra A, Gosain S, Young KM (1999) IT-intensive value innovation in the electronic economy: insights from Marshall Industries. MIS Q 23(3):305–335

Elango B (2009) Impact of insurers' product variety on performance across underwriting cycles. Manage Decis 47(2):359–374

Elbashir M, Collier PA, Sutton SG, Hunton J, Vaassen E (2008) Business intelligence systems use to leverage enterprise-wide accounting information in shared data environments. Paper presented at the European conference on accounting information systems (ECAIS) 2008, Maastricht, The Netherlands

Enabled O (2003) ILL Number: 26371694. Month, 31

Erber F (1999) O sistema de inovações em uma economia monetária: uma agenda de pesquisas. Globalização & Inovação Localizada. Experiências de Sistemas Locals no Mercosul, IBICT/MCT, Brasília

Eriksson K, Johanson J, Majkgårw A, Sharma DD (2000) Effect of variation on knowledge accumulation in the internationalization process. International Studies of Management & Organization 30(1): 26

Estrada I, Fuente G, Martín-Cruz N, Esgueva AV, No, P., & No, F. Technological joint venture formation under the real options approach

Farjoun M (2002) Towards an organic perspective on strategy. Strateg Manage J 23(7):561–594

Fellenz M (2000) Flexibility in management theory: towards clarification of an elusive concept. Available at SSRN 1098111

Fichman RG (2004) Real options and IT platform adoption: implications for theory and practice. Inf Syst Res 15(2):132–154

Fink L, Neumann S (2009) Taking the high road to web services implementation: an exploratory investigation of the organizational impacts. ACM SIGMIS Database 40(3):84–108

Finkelstein S, Sanford SH (2000) Learning from corporate mistakes: the rise and fall of iridium. Organ Dyn 29(2):138–148

Fisch JH (2008) Internalization and internationalization under competing real options. J Int Manage 14(2):108–123

Fisch JH. Papers in global business management

Fisch JH, Roß JM (2008) Optionen der Internationalisierung offenhalten oder neue eröffnen-Panelstudie über den Markteintrittszeitpunkt deutscher Investoren im Ausland. Ausländische Direktinvestitionen: Neuere Entwicklungen, Entscheidungsinstrumente und führungsrelevante Folgen, 39

Floricel S, Michela JL (2007) Dynamique environnementale, renouvellement stratégique et avantage concurrentiel. XVIème Conférence Internationale de Management Stratégique. Montreal, Canada

Folta TB, Ferrier WJ (1996) International expansion through sequential investment: the effects of national culture on buyouts and dissolutions in biotechnology partnerships. Available at SSRN 43561

Ford DN, Sobek D (2005) Modeling real options to switch among alternatives in product development. IEEE Trans Eng Manage 52(2):1–11

Ford D, Lander D, Voyer J (2004) Business strategy and real options in the context of large engineering projects. J Compet

Foss NJ (1998) Real options and the theory of the firm: Institute for Industriøkonomi og Virksomhedsstrategi, Handelshøjskolen i København

From S (2000) Uncertainty and new venture investments: some empirical evidence from young Italian firms. Int J Entrep 4:1

Fuentelsaz L, Gomez J, Polo Y (2002) Followers' entry timing: evidence from the Spanish banking sector after deregulation. Strateg Manage J 23(3):245–262

für die wertorientierte Unternehmensführung I, Hungenberg H, Wulf T. Einsatzfelder und Operationalisierung der Realoptionstheorie

Gavetti G, Levinthal D (2000) Looking forward and looking backward: cognitive and experiential search. Adm Sci Q 45(1):113–137
Gibson CB, Manuel JA (2003) Building trust. Virtual teams that work. San Francisco: Jossey-Bass, 59–86
Gielens K, Dekimpe M (2004) How to seize a window of opportunity: the entry strategy of retail firms into transition economies. Available at SSRN http://ssrn.com/abstract=595006
Giere J (2007) Marketingflexibilität: Eine empirische Analyse ihrer Konzeptionalisierung, Operationalisierung und Erfolgswirkung: Duv
Ginogluc D. Top management staffing in international strategic alliances: a conceptual explanation of decision perspective and objective formation. International Business Review. Special Issue Advances in International Marketing Theory and Practice, 173
Gosain S (2007) Realizing the vision for web services: strategies for dealing with imperfect standards. Inf Syst Front 9(1):53–67
Greenley GE, Oktemgil M (1998) A comparison of slack resources in high and low performing British companies. J Manage Stud 35(3):377–398
Greenley G, Hooley G, Saunders J (2004) Management processes in marketing planning. Eur J Mark 38(8):933–955
Grewal R, Chakravarty A, Ding M, Liechty J (2008) Counting chickens before the eggs hatch: on the valuation of new drug (product) development portfolios in the pharmaceutical sector. Institute for the Study of Business Markets, The Pennsylvania State University, University Park, PA
Grewal R, Tansuhaj P (2001) Building organizational capabilities for managing economic crisis: the role of market orientation and strategic flexibility. Journal of Marketing 65(2):67–80
Grewal R, Johnson JL, Sarker S (2007) Crises in business markets: implications for interfirm linkages. J Acad Mark Sci 35(3):398–416
Grimpe C, Hussinger K. (2008) Building and blocking: the two faces of technology acquisition. ZEW Discussion Papers No. 08-042
Grimpe C, Hussinger K (2008) Rent appropriation and competitor blocking: the two faces of technology acquisition DRUID 25th Celebration Conference 2008, Frederiksberg, Denmark
Grimpe C, Hussinger K, Statistieken SOO (2008) Building and blocking: the two faces of technology acquisition. ZEW Discussion Papers No. 08-042
Grimpe C, Hussinger K, Statistieken SOO (2007) Firm acquisitions and technology strategy: corporate versus private equity investors, ZEW Discussion Papers No. 07-066
Grimpe C, Kaiser U, Porcelaenshaven B (2008) Gains and pains from contract research: a transaction and firm-level perspective, ZEW Discussion Papers No. 08-002
Grimpe C, Hussinger K (2008) Market and technology access through firm acquisitions: beyond one size fits all. ZEW, Zentrum für Europ. Wirtschaftsforschung
Grimpe C, Sofka W (2009) Search patterns and absorptive capacity: low- and high-technology sectors in European countries. Res Policy 38(3):495–506
Guler I (2007a) Throwing good money after bad? Political and institutional influences on sequential decision making in the venture capital industry. Adm Sci Q 52(2):248–285
Guler I (2007b) An empirical examination of management of real options in the US venture capital industry. Adv Strateg Manage 24:485–506
Gunther McGrath R (1997) Falling forward: real options reasoning and entrepreneurial failure. Acad Manage Rev 24(1):13–30
Gunther McGrath R, Nerkar A (2004) Real options reasoning and a new look at the R&D investment strategies of pharmaceutical firms. Strateg Manage J 25(1)
Haas F (2004) Effizienztreiber innovativer Prozesse: Anwendung der Data Envelopment Analysis am Beispiel der elektronischen C-teile-beschaffung: Duv
Hackett SM (2004) Real options and the option to incubate: an exploratory study of the process of business incubation. Available at SSRN 1260438
Hackett SM, Dilts DM (2004) A real options-driven theory of business incubation. J Technol Transf 29(1):41–54

Hagedoorn J, Sadowski B (1999) The transition from strategic technology alliances to mergers and acquisitions: an exploratory study. J Manage Stud 36(1):87–107

Hamsal M, Agung IGN (2007) Paradoxical strategies and firm performance: the case of Indonesian banking industry. South East Asian Journal of Management, Vol 1(1):43–64

Hariharan S, Brush TH (1999) Plant scale in entry decisions: a comparison of start-ups and established firm entrants. Manag Decis Econ 20:353–364

Helfat C, Bowman EH (2001) Comment: Edward H. Bowman 1925–1998. Strateg Manage J 22(1):25–26

Helfat CE, Raubitschek RS (2000) Product sequencing: co-evolution of knowledge, capabilities and products. Strateg Manage J 21:961–979

Hertzel M, Smith JK, Smith RL (2001) Competitive impact of strategic restructuring: evidence from the telecommunications industry, vol 10. Oxford University Press, Oxford

Hienerth C (2004) Theorie und Realität unternehmerischer Strategien: Eine Anwendung des Konfigurationsansatzes in Kleinbetrieben: Dt. Univ.-Verl

Hienerth C, Kessler A (2006) Measuring success in family businesses: the concept of configurational fit. Fam Bus Rev 19(2):115

Hilmola OP, Torkkeli M, Savolainen VV (2009) Intangibles and business performance–a technical efficiency approach. Int J Entrep Innov Manage 9(4):364–379

Homburg C, Hoyer WD, Fassnacht M (2002) Service orientation of a retailer's business strategy: dimensions, antecedents, and performance outcomes. J Mark 66(4):86–101

Hommel U, Scholich M, Vollrath R (eds) (2001) Realoptionen in der Unternehmenspraxis: Wert schaffen durch Flexibilität. Springer, Berlin/Heidelberg/New York

Hopkins HD, Swift T (2008) Business leaders speak out: their real strategic problems. J Bus Strategy 29(5):32–37

Hovhannisian K (2000) Application of the real-option technique to investments in learning http://www.druid.dk/uploads/tx_picturedb/dw2001-392.pdf

Hundsnes T (2000) The management of complexity SNF-Report No. 58/2000

Hussinger K, Grimpe C (2007) Firm acquisitions and technology strategy: corporate versus private equity investors, ZEW Discussion Papers No. 07-066

Hussinger K, Grimpe C (2009) Market and technology access through firm acquisitions: beyond one size fits all. Progress In International Business Research vol 3

Husted BW (2005) Risk management, real options, and corporate social responsibility. J Bus Ethics 60(2):175–183

Industry C (2005) Selection in strategic alliance activity. Eur Manage J 23(4):361–381

Jarvenpaa SL, Tiller EH (1999) Integrating market, technology, and policy opportunities in e-business strategy. J Strateg Inf Syst 8(3):235–249

Jarzabkowski P (2003) Relevance in theory and relevance in practice: strategy theory in practice. Paper presented at the 19th EGOS Colloquium, Copenhagen, Denmark

Jarzabkowski P, Wilson DC (2006) Actionable strategy knowledge: a practice perspective. Eur Manage J 24(5):348–367

Jasinski DW, Huff AS (2002) Knowledge-based system to study strategic options. In Mapping Strategic Knowledge. Huff, A. S., & Jenkins, M. (Eds.). Sage Publications Limited. Los Angeles, CA

Jensen BK, Cline M, Guynes CS (2007) HIPPA, privacy and organizational change: a challenge for management. ACM SIGCAS Computers and Society 37(1):12–17

Jiang B, Yao T, Feng B (2008) Valuate outsourcing contracts from vendors' perspective: a real options approach. Decis Sci 39(3):383–405

Jing L, Dhanaraj C, Shockley RL (2008) Joint venture evolution: extending the real options approach. Manag Decis Econ 29(4):317–336

Johnson JL, Lee RPW, Saini A, Grohmann B (2003) Market-focused strategic flexibility: conceptual advances and an integrative model. J Acad Mark Sci 31(1):74–89

Jun L, Weiguo Z, Bingshan W (2007) Study on the asymmetric information problems in knowledge alliance. Paper presented at the 2007 IEEE international conference on automation and logistics, Jinan, Shandong, China

Kellermanns F, Floyd S (2005) The effect of strategic consensus on organizational flexibility. Erfolgsfaktor Flexibilität: Strategien und Konzepte für wandlungsfähige Unternehmen, 55

Ketchen Jr. DJ, Ireland RD, Snow CC (2007) Strategic entrepreneurship, collaborative innovation, and wealth creation. Strateg Entrep J Vol 1 (3–4). Pages 371–385

Ketkar S, Sett PK (2009) HR flexibility and firm performance: analysis of a multi-level causal model. Int J Hum Res Manage 20(5):1009–1038

Klingebiel R, Street T. These papers are produced by Judge Business School, University of Cambridge. They are circulated for discussion purposes only. Their contents should be considered preliminary and are not to be quoted without the authors' permission

Klossek A (2008) Market entry and expansion through international joint ventures: a multi-causal analysis of international joint venture performance. Universitätsbibliothek der TU BAF

Knoll S (2008) Cross-business synergies: a typology of cross-business synergies and a mid-range theory of continuous growth synergy realization. Gabler, Betriebswirt.-Vlg

Kogut B, Kulatilaka N (2001) Capabilities as real options. Organ Sci 12(6):744–758

Kogut B, Kunreuther H (2001) Introduction to focused issue: risk, managers, and options in organizations. Organ Sci 12(5):579–581

Kraatz MS, Zajac EJ (2001) How organizational resources affect strategic change and performance in Turbulent environments: theory and evidence. Organ Sci 12(5):632–657

Kramer M, Müller D (2004) Realoptionsmodelle als Instrumente des Investitionscontrollings von Umweltinnovationen. Nachhaltiges Innovations management: Heinz Strebel zum 65. Geburtstag 275

Kranz M (2007) Management von Strategieprozessen: Von der strategischen Planung zur integrierten Strategieentwicklung: Duv

Kremljak Z, Sakic N (2006) Integral uncertainty estimate of production capabilities development. Transactions of FAMENA 30(2):61–68

Kremljak Z, Polajnar A, Buchmeister B (2005) A heuristic model for the development of production capabilities. Strojniski Vestnik 51(11):674–691

Kukovetz K (2002) Decision-making processes in emerging markets. Unpublished Doctoral Dissertation, University of St. Gallen, Switzerland

Kumar MVS (2005) The value from acquiring and divesting a joint venture: a real options approach. Strateg Manage J 26(4):321–331

Kuruzovich JN (2006) Electronic mediation, transformation, and business value: three essays in the retail auto industry Dissertation. University of Maryland, College Park, MD

Kyläheiko K, Sandström J (2007) Strategic options-based framework for management of dynamic capabilities in manufacturing firms. Management 18(8):966–984

Kylaheiko K, Jantunen A, Sandstrom J (2008) Valuing technological capabilities? The strategic options perspective. Int J Bus Excell 1(1):71–91

LA I, Empresarial D, Como G, Moderadoras V. Facultad de Ciencias Económicas y Empresariales Departamento de Organización de Empresas

Laaksonen P (2005) Managing strategic change: new business models applying wireless technology as a source of competitive edge. Acta Universitatis Lappeenrantaensis. Lappeenranta University of Technology, Lappeenranta, Finland

Laamanen T (1999) Option nature of company acquisitions motivated by competence acquisition. Small Bus Econ 12(2):149–168

Labriola F (2006) Ganzheitliches time-to-market management: Planung und Umsetzung von Produktinnovationen unter besonderer Berücksichtigung des Wettbewerbsfaktors Zeit. Universitätsbibliothek der TU BAF

Lado AA, Boyd NG, Hanlon SC (1997) Competition, cooperation, and the search for economic rents: a syncretic model. Acad Manage Rev 22(1):110–141

Lane PJ, Koka BR, Pathak S (2006) The reification of absorptive capacity: a critical review and rejuvenation of the construct. Acad Manage Rev 31(4):833–863

Lankton N, Luft J (2008) Uncertainty and industry structure effects on managerial intuition about information technology real options. J Manage Inf Syst 25(2):203–240

Lashley K (2004) The effects of university technology commercialization practices on licensing income-a comparative study Master's Thesis. Vanderbuilt University, Nashville, TN
Latham SF, Braun M (2009) Managerial risk, innovation, and organizational decline. J Manage 35(2):258
Laverty KJ (1996) Economic “short-termism”: the debate, the unresolved issues, and the implications for management practice and research. Acad Manage Rev 21(3):825–860
Lee GK (2008) Relevance of organizational capabilities and its dynamics: what to learn from entrants' product portfolios about the determinants of entry timing. Strateg Manage J 29(12)
Lee SH, Chung CC (2007) 2. Geographic diversification: risk reduction or operational flexibility. A New Generation in International Strategic Management, 20
Lee SH, Makhija M (2009) Flexibility in internationalization: is it valuable during an economic crisis? Strateg Manage J 30(5):537–555
Lee GK, Paruchuri S (2008) Entry into emergent and uncertain product-markets: the role of associative rhetoric. Acad Manage J (AMJ) 51(6):1171–1188
Lee CS, Vonortas NS (2002) Toward an integrated model of strategy formulation for strategic technical alliances. Int J Technol Transf Commer 1(3):292–312
Lee S, Peng MW, Barney JB (2007) Bankruptcy law and entrepreneurship development: a real options perspective. Acad Manage Rev 32(1):257
Lee SH, Makhija M, Paik Y (2008) The value of real options investments under abnormal uncertainty: the case of the Korean economic crisis. J World Bus 43(1):16–34
Lee SH, Beamish PW, Lee HU, Park JH (2009) Strategic choice during economic crisis: domestic market position, organizational capabilities and export flexibility. J World Bus 44(1):1–15
Lee S-H, Makhija M, Paik Y (2011) The value of real options investments under abnormal uncertainty: the case of the Korean economic crisis. J World Bus 43(1):16–34
Lei DT (1997) Competence-building, technology fusion and competitive advantage: the key roles of organisational learning and strategic alliances. Int J Technol Manage 14(2):208–237
Lei DT (2000) Industry evolution and competence development: the imperatives of technological convergence. Int J Technol Manage 19(7):699–738
Lei D (2003) Competition, cooperation and learning: the new dynamics of strategy and organisation design for the innovation net. Int J Technol Manage 26(7):694–716
Lei D, Hitt MA, Goldhar JD (1996) Advanced manufacturing technology: organizational design and strategic flexibility. Organ Stud (Walter de Gruyter GmbH & Co KG) 17(3):501
Leiblein MJ (2003) The choice of organizational governance form and performance: predictions from transaction cost, resource-based, and real options theories. J Manage 29(6):937
Leiblein MJ, Miller DJ (2003) An empirical examination of transaction- and firm-level influences on the vertical boundaries of the firm. Strateg Manage J 24(9):839–859
Leitmannslehner S (2008) Dynamics of networks: from franchising to licensing case study Getitfix. Doctoral Dissertation. University of Vienna, Vienna, Austria
Li J (2007a) Real options theory and international strategy: a critical review. Adv Strateg Manage 24:71–105
Li PP (2007b) Social tie, social capital, and social behavior: toward an integrative model of informal exchange. Asia Pac J Manage 24(2):227–246
Li X (2009) The nature of the firm: a relationship-based theory. Working Paper wp3-2009-xl, Department of International Economics and Management, Copenhagen Business School, Denmark
Li J, Tong TW (2007) 6. Real options theory and international strategic management. A New Generation in International Strategic Management 100
Li Y, James B, Madhavan R, Mahoney JT (2007) Real options: taking stock and looking ahead. Adv Strateg Manage 24:33–66
Li J, Dhanaraj C, Shockley RL (2008) Joint venture evolution: extending the real options approach. Manage Decis Econ 29(4)
Lim T, Lo AW, Merton RC, Scholes MS (2006) The Derivatives Sourcebook: Now Pub
Lint, O., & Pennings, E. Eindhoven Centre for Innovation Studies, The Netherlands.
Lint O. Retrospective insights from real options in R&D. Eindhoven Center for Innovation Studies

Lint O, Pennings E (1999) Finance and strategy: time-to-wait or time-to-market? Long Range Plann 32(5):483–493
Lloréns FJ, Molina LM, Verdú AJ (2005) Flexibility of manufacturing systems, strategic change and performance. Int J Prod Econ 98(3):273–289
Loranth G, Street T. These papers are produced by Judge Business School, University of Cambridge. They are circulated for discussion purposes only. Their contents should be considered preliminary and are not to be quoted without the authors' permission
Luo Y, Peng MW (1999) Learning to compete in a transition economy: experience, environment, and performance. J Int Bus Stud 30(2):269–271
M'Chirgui Z, Chanel O (2007) La dynamique des alliances et des acquisitions dans l'industrie de la carte a puce. Economie Appliquee 60(4):107–140
Ma Q, Wang Y (2006) Entrepreneuring on Website. Paper presented at the 2006 IEEE international conference on management of innovation and technology, Singapore
MacLeod N, Agriculture CT. Case study 4 effective strategies for increasing the suitability and adoption of complex technologies for sustainable grazing land management. Case Studies in Increasing the Adoption of Sustainable Resource Management Practices, 183
Macmillan I, McGrath R (1996) Technology positioning investments: a logic for real options strikes. Wharton School of the University of Pennsylvania, Snider Entrepreneurial Center. University Park, PA
Mahoney JT (2005) Economic foundations of strategy. Sage Publications. Thousand Oaks, CA
Mahoney JT, McGahan AM (2007) Strategic organization. Strateg Organ 5(1):79–99
Mahoney JT, Sanchez R (2004) Building new management theory by integrating processes and products of thought. J Manage Inq 13(1):34
Malmström M, Wincent J. Competence acquisition in small firms. Paper Presented at the 1st Nordic Innovation Research Conference – Finnkampen
Malos S (2010) Regulatory effects and strategic global staffing profiles: beyond cost concerns in evaluating offshore location attractiveness. Empl Responsib Rights J Vol 22(2):1–19
Malos SB, Campion MA (1995) An options-based model of career mobility in professional service firms. Acad Manage Rev 20(3):611–644
Marino L, Strandholm K, Steensma HK, Weaver KM (2002) The moderating effect of national culture on the relationship between entrepreneurial orientation and strategic alliance portfolio extensiveness. Entrep Theory Pract 26(4):145
Martin JA, Eisenhardt KM (2003) Cross-business synergy: recombination, modularity and the multi-business team. Academy of Management 2003 Best Paper Proceedings
Mathews JA (2003) Strategizing by firms in the presence of markets for resources. Ind Corp Change 12(6):1157–1193
Matthyssens P, Pauwels P, Vandenbempt K (2005) Strategic flexibility, rigidity and barriers to the development of absorptive capacity in business markets: themes and research perspectives. Ind Mark Manage 34(6):547–554
McCarthy IP (2004) Manufacturing strategy: understanding the fitness landscape. Int J Oper Prod Manage 24(1):124–150
McGill JP, Santoro MD (2004) Managing value and vulnerability in complex collaborations: allying with competitors. In: Complex collaboration: building the capabilities for working across boundaries. Michael Martin Beyerlein, Douglas A. Johnson, Susan T. Beyerlein (Eds) p 263. Elsevier, San Diego, CA
McGill, Joseph P., and Michael D. Santoro (2004) "Managing value and vulnerability in complex collaborations: allying with competitors." In complex collaboration: building the capabilities for working across boundaries (Vol. 10). Beyerlein, M. M., Johnson, D. A., & Beyerlein, S. T. (Eds.). 263–286. Emerald Group Publishing. Elsevier, Oxford, UK
McGrath R (1996) The trouble with competence: opportunities and limitations in the adolescence of the resource-based view. Annual Academy of Management Meetings, Business Policy and Strategy Division, Cincinnati. Ohio
McGrath RG (1997) A real options logic for initiating technology positioning investments. Acad Manage Rev 22(4):974–996

McGrath RG (1999) Falling forward: real options reasoning and entrepreneurial failure. Acad Manage Rev 24:13–30
McGrath RG (2001) Exploratory learning, innovative capacity, and managerial oversight. Acad Manage J 44(1):118–131
McGrath RG, Boisot M (2003) Real options reasoning and the dynamic organization: strategic insights from the biological analogy. Leading and Managing People in the Dynamic Organization, 201
McGrath RG, MacMillan IC (2000) Assessing technology projects using real options reasoning. Res Technol Manage 43(4):35–49
McGrath RG, Nerkar A (2004) Real options reasoning and a new look at the R&D investment strategies of pharmaceutical firms. Strateg Manage J 25(1):1–21
McGrath RG, Macmillan IC, Venkataraman S (1995) Defining and developing competence: a strategic process paradigm. Strateg Manage J 16:251–275
McGrath RG, Chen MJ, MacMillan IC (1998a) Multimarket maneuvering in uncertain spheres of influence: resource diversion strategies. Acad Manage Rev 23:724–740
McGrath RG, Ming-Jer C, MacMillan IC (1998b) Multimarket maneuvering in uncertain spheres of influence: resource diversion strategies. Acad Manage Rev 23(4):724–740
McGrath RG, Ferrier WJ, Mendelow AL (2004) Real options as engines of choice and heterogeneity. Acad Manage Rev 29(1):86–101
McGuire J, Dow S (2003) The persistence and implications of Japanese Keiretsu organization. J Int Bus Stud 34(4):374–389
Meier H, Christofides N, Salkin G (2001) Capital budgeting under uncertainty–an integrated approach using contingent claims analysis and integer programming. Oper Res 49(2):196–206
Menor LJ, Roth AV, Mason CH (2001) Agility in retail banking: a numerical taxonomy of strategic service groups. Manuf Serv Oper Manage 3(4):273
Merton RC (1998) Applications of option-pricing theory: twenty-five years later. Am Econ Rev 88(3):323–349
Miller KD (1998) Economic exposure and integrated risk management. Strateg Manage J 19(5):497–514
Miller KD, Reuer JJ (1996) Measuring organizational downside risk. Strateg Manage J 17(9):671–691
Miller KD, Shapira Z (2004) An empirical test of heuristics and biases affecting real option valuation. Strateg Manage J 25(3):269–284
Miller KD, Waller HG (2003) Scenarios, real options and integrated risk management. Long Range Plann 36(1):93–107
Mölls S (2004) Grenzpreisbestimmung von Unternehmensnetzwerken: Netzwerkeffekte in der Unternehmensbewertung: Duv
Moitra D (2008) Globalization of R&D: leveraging offshoring for innovative capability and organizational flexibility. Doctoral dissertation, Erasmus University Rotterdam, Rotterdam, The Netherlands
Moon H, Conlon DE, Humphrey SE, Quigley N, Devers CE, Nowakowski JM (2003) Group decision process and incrementalism in organizational decision making. Organ Behav Hum Decis Process 92(1/2):67
Mosakowski E (1997) Strategy making under causal ambiguity: conceptual issues and empirical evidence. Organ Sci 8(4):414–442
Mosakowski E, Earley PC (2000) A selective review of time assumptions in strategy research. Acad Manage Rev 25(4):796–812
Mühlfeld KS (2004) Strategic Shifts Between Business Types: A Transaction Cost Theory-based Approach Supported by Dyad Simulation: Duv
Muramatsu N, Lee SYD, Alexander JA (2000) Hospital provision of institutional long-term care pattern and correlates. Gerontologist 40(5):557–567
Namuduri S (2006) Analysis of resource-sharing decisions in dyadic collaborative knowledge creation: a game-theoretic approach, Doctoral Dissertation. Managerial Sciences Dissertations. Paper 9. Georgia State University, Atlanta, GA http://digitalarchive.gsu.edu/managerialsci_diss/9

Narayanan VK, Pinches GE, Kelm KM, Lander DM (2000) The influence of voluntarily disclosed qualitative information. Strateg Manage J 21(7):707

Natarajan VS, Munch J (2007) Viewing new product development through the real options lens: an empirical investigation of the market reaction and the role of contextual factors. Doctoral Dissertation, University of Texas at Arlington, Arlington, TX

Ndofor HA, Levitas E (2004) Signaling the strategic value of knowledge. J Manage 30(5):685

Ng DW (2007) A modern resource based approach to unrelated diversification. J Manage Stud 44(8):1481–1502

Nippa M, Petzold K (2000) Ökonomische Erklärungs-und Gestaltungsbeiträge des Realoptionen-Ansatzes: Bergakad., Faculty of Economics and Business Administration

Nippa M, Petzold K (2003) Zur Anwendbarkeit des Realoptionenansatzes als Instrument zur Unterstützung strategischer Entscheidungsprozesse-Indizien kontingenz-theoretischer Bewertungsnotwendigkeiten. Strategische Prozesse und Pfade, 151

Omar A El Sawy, Arvind Malhotra, Sanjay Gosain, and Kerry M. Young (1999) IT-Intensive value innovation in the electronic economy: insights from marshall industries. MIS Quarterly 23(3):303–335

Oriani R, Sobrero M. Market valuation of firms' technological knowledge: a real options perspective. Paper presented at the international conference of the strategic management society

Oriani R, Sobrero M (2002) Assessing the market valuation of firms' technological knowledge using a real options perspective. Paper presented at the annual meeting of the academy of management, Denver, CO

Oriani R, Sobrero M (2008) Uncertainty and the market valuation of R&D within a real options logic. Strateg Manage J 29(4):343–361

Ortiz DC (2009) The impact of dominant logic orientation (exploitation Vs exploration) on the firm's real options recognition. Decision Sciences Institute Paper http://www.swdsi.org/swdsi2009/Papers/9S06.pdf

Pan Y, Chi PSK (1999) Financial performance and survival of multinational corporations in China. Strateg Manage J 20:359–374

Pan Y, Li S, Tse DK (1999) The impact of order and mode of market entry on profitability and market share. J Int Bus Stud 30(1):81–83

Pandza K, Horsburgh S, Gorton K, Polajnar A (2003) A real options approach to managing resources and capabilities. Int J Oper Prod Manage 23(9):1010

Pandza K, Horsburgh S, Gorton K, Polajnar A (2003) A real options approach to managing resources and capabilities. International Journal of Operations & Production Management, 23(9):1010–1032

Parayre R, Hurry D (2001) Corporate investment and strategic stability in hypercompetition. Manage Decis Econ 22:281–298

Park SH, Kim D (1997) Market valuation of joint ventures: joint venture characteristics and wealth gains. J Bus Ventur 12(2):83–108

Park SH, Ungson GR (1997) The effect of national culture, organizational complementarity, and economic motivation on joint venture dissolution. Acad Manage J 40(2):279–307

Parmigiani AE (2007) Why do firms both make and buy? An investigation of concurrent sourcing. Strateg Manage J 28(3):285–311

Partnerships B, Folta T, Ferrier W. International expansion through sequential investment: the effects of national culture on buyouts and dissolutions in biotechnology partnerships

Pena I (2009) Interpreneurial strategic behaviour and business performance. Int J Entrep Innov Manage 10(1):3–19

Peña I, Miner J, Otamendi N. Opciones Reales: Aplicación a Decisiones de Internacionalización

Perez M. La gestion des alliances autour de projets innovants: une approche par les options réelles

Perez M, Delerue H (2009) Unilateral commitment in alliances: an optional behaviour. J Manage Dev 28(2):134–149

Pillai KG, Goldsmith RE (2006) Calibrating managerial knowledge of customer feedback measures: a conceptual model. Mark Theory 6(2):223

Pontoppidan AIC, Indledning A. Risikoledelse set i et viden-og værdiskabende ledelsesperspektiv–når ledelsen bliver nødt til at udforske det strategiske mulighedsrum

Power B (2008) Real options reasoning as a tool for managerial decision making. Encyclopedia of Decision Making and Decision Support Technologies, 766

Power B, Reid GC (2005) Flexibility, firm-specific turbulence and the performance of the long-lived small firm. Rev Ind Organ 26(4):415–443

Pynnönen M (2004) Mobile E-commerce business model-a value web based approach to business models in mobile gaming industry

Reid GC (2007) Risk appraisal and venture capital in high technology new ventures. Routledge, New York, NY

Reuer J (2002) How real are real options? The case of international joint ventures. Creating value: Winners in the new business environment, 61–84

Reuer JJ, Tong TW (2003) Switching options and coordination costs in multinational firms. Proceedings of the 4th IGMS Ciber Research forum, Temple University, Philadelphia, USA

Reuer JJ, Leiblein MJ (2000) Downside risk implications of multinationality and international joint ventures. Acad Manage J 43(2):203–214

Reuer JJ, Tong TW (2005) Real options in international joint ventures. J Manage 31(3):403

REUER Jeffrey J, TONG Tony W (2007) Real options theory. Advances in Strategic Management vol 24

Reuer JJ, Tong TW (2007) Corporate investments and growth options. Manage Decis Econ 28(8):863–877

Rindfleisch H, Heine K. Modern governance: Der Beitrag der neueren Organisationsforschung. Governance-Theorien oder Governance als Theorie? 211

Rittippant N, Rasheed A (2007) Multinational strategies as options creation and exercise: an analysis of country, industry, and firm specific characteristics

Rivoli P, Salorio E (1996) Foreign direct investment and investment under uncertainty. J Int Bus Stud 27(2):335–357

Robson MJ, Paparoidamis N, Ginoglu D (2003) Top management staffing in international strategic alliances: a conceptual explanation of decision perspective and objective formation. Int Bus Rev 12(2):173

Saad M, Zantout Z (2009) Stock price and systematic risk effects of discontinuation of corporate R&D programs. J Empir Financ 16(4):568–581

Sahaym A, Steensma HK, Schilling MA (2007) The influence of information technology on the use of loosely coupled organizational forms: an industry-level analysis. Organ Sci 18(5):865–880

Sallleh NAM, Jusoh R, Isa CR. The effect of information systems sophistication on use of performance measures: a study on Malaysian financial services institutions

Samaddar S, Kadiyala SS (2006) An analysis of interorganizational resource sharing decisions in collaborative knowledge creation. Eur J Oper Res 170(1):192–210

Sambamurthy V, Bharadwaj A, Grover V (2003) Shaping agility through digital options: reconceptualizing the role of information technology in contemporary firms. MIS Q 27(2):237–263

Sandri S, Schade C, Mußhoff O, Odening M (2009) Holding On for Too Long? An Experimental Study on Inertia in Entrepreneurs' and Non-Entrepreneurs' Disinvestment Choices SiAg-Working Paper 02

Sapsed J (2002) Visions, competences and rational strategizing: some theory and evidence

Scherpereel CM (2008) The option-creating institution: a real options perspective on economic organization. Strateg Manage J 29(5):455–470

Schilling MA (1998) Technological lockout: an integrative model of the economic and strategic factors driving technology success and failure. Acad Manage Rev 23(2):267–284

Schilling MA (2002) Technology success and failure in winner-take-all markets: the impact of learning orientation, timing, and network externalities. Acad Manage J 45(2):387–398

Schilling MA, Hill CWL (1998a) Managing the new product development process: strategic imperatives. Acad Manage Exec 12(3):67–81

Schilling MA, Hill CWL (1998b) Managing the new product development process: strategic imperatives. Acad Manage Exec 1993–2005:67–81

Schmidt T, Schwiebacher F, Sofka W. The effects of experience on selecting innovation projects-better the devil you know

Schminke M (2001) Considering the business in business ethics: an exploratory study of the influence of organizational size and structure on individual ethical predispositions. J Bus Ethics 30(4):375–390

Schneider A (2009) Die Rolle flexibler Organisationsstrukturen im Innovationsmanagement. GRIN Verlag

Schoemaker PJH (2008) The future challenges of business: rethinking management education. Calif Manage Rev 50(3):19–139

Schwarz A, Mehta M, Johnson N, Chin WW (2007) Understanding frameworks and reviews: a commentary to assist us in moving our field forward by analyzing our past. ACM SIGMIS Database Newsletter. Volume 38 Issue 3, August 2007. Pages 29–50. Publisher ACM New York, NY, USA

Sea Jin C (1995) International expansion strategy of Japanese firms: capability building through sequential entry. Acad Manage J 38(2):383–407

Sea-Jin C, Rosenzweig PM (1998) Industry and Regional Patterns in Sequential Foreign Market Entry. J Manage Stud 35(6):797–822

Segelod E (2001) Learning through grafting in ventures in new areas. Scand J Manage 17(3):305–328

Seung-Hyun L, Peng MW, Barney JB (2007) Bankruptcy law and entrepreneurship development: a real options perspective. Acad Manage Rev 32(1):257–272

Sharma DD, Blomstermo A (2003) A critical review of time in the internationalization process of firms. J Glob Mark 16(4):53–71

Sheasley WD (2000) Taking an options approach to new technology development. Res Technol Manage 43(6):37–43

Shepherd DA, Zacharakis A (2001) The venture capitalist-entrepreneur relationship: control, trust and confidence in co-operative behaviour. Venture Capital 3(2):129–149

Shimizu K, Hitt MA (2004) Strategic flexibility: organizational preparedness to reverse ineffective strategic decisions. Acad Manage Exec 18(4):44–59

Shyam Kumar MV (2005) The value from acquiring and divesting a joint venture: a real options approach. Strateg Manage J 26(4)

Sirmon DG, Hitt MA, Ireland RD (2007) Managing firm resources in dynamic environments to create value: looking inside the black box. Acad Manage Rev 32(1):273–292

Skilton PF (2003) Portable integration expertise as the foundation of flexible integration capability. Int J Organ Anal 11(4):319–338

Slater SF, Reddy VK, Zwirlein TJ (1998) Evaluating strategic investments complementing discounted cash flow analysis with options analysis. Ind Mark Manage 27(5):447–458

Smit HTJ, Trigeorgis L (2007) Strategic options and games in analysing dynamic technology investments. Long Range Plan 40(1):84–114

Smit H, Trigeorgis L (2006) Strategic planning: valuing and managing portfolios of real options. R&D Manage 36(4):403–419

Sofka W, Grimpe C (2008) Managing search strategies for open innovation-the role of environmental munificence as well as internal and external R&D, ZEW Discussion Papers No. 08-075

Sorescu AB, Chandy RK, Prabhu JC (2003) Sources and financial consequences of radical innovation: insights from pharmaceuticals. J Mark 67(4):82–102

Srinivasan R (2006) Dual distribution and intangible firm value: franchising in restaurant chains. J Mark 70(3):120–135

Srinivasan R. (2006) Dual distribution and intangible firm value: franchising in restaurant chains. J Mark 70(3):120–135

Starbuck WH, Barnett ML, Baumard P (2008) Payoffs and pitfalls of strategic learning. J Econ Behav Organ 66(1):7–21

Statistic F (2006) Business start-up process differences reported by novice, serial, and parallel founders. New firm startups, 448

Steensma HK, Fairbank JF (1999) Internalizing external technology: a model of governance mode choice and an empirical assessment. J High Technol Manage Res 10(1):1–35

Su N, Akkiraju R, Nayak N, Goodwin R (2009) Shared services transformation: conceptualization and valuation from the perspective of real options. Decis Sci 40(3):381–402

Tailan C (2000) Option to acquire or divest a joint venture. Strateg Manage J 21(6):665

Talay MB, Akdeniz MB (2009) What causes break-ups? Factors driving the dissolution of marketing-oriented international joint ventures. In New Challenges to International Marketing, Rudolf R. Sinkovics, Pervez N. Ghauri (Eds). JAI Press, Emerald Group Publishing Limited, Bingley, UK

Tang X (2007) Inter-organizational relationship portfolio management: a digital enablement perspective of process alignment and process innovation

Tang X (2007) Inter-organizational Relationship Portfolio Management: A Digital Enablement of Perspective of Process Alignment and Process Innovativeness. Doctoral Dissertation. Georgia State University, Atlanta, Georgia

Tang Z, Ji D (2005) Growth option behind patenting behavior. Paper presented at the 2005 IEEE international engineering management conference, St. John's, Newfoundland & Labrador, Canada

Templeton G, Snyder C (1999) A model of organisational learning based on control. Int J Technol Manage 18(5):705–719

Templeton GF, Snyder CA (2000) Precursors, contexts and consequences of organisational learning. Int J Technol Manage 20(5):765–781

Templeton GF, Lewis BR, Snyder CA (2002) Development of a measure for the organizational learning construct. J Manage Inf Syst 19(2):175–218

Thomas AR (2007) Entry strategy concepts, determinants and options of US firms into Romania. Manage Mark (3):3

Thurner MO (2003) Are Real Options Dead? Considerations for Theory and Practice. Research Institute for International Management, University of St. Gallen

Tian J, Wang K, Chen Y, Johansson B (2010) From IT deployment capabilities to competitive advantage: an exploratory study in China. Inf Syst Front Vol 12(3):239–255

Tong W (2004) Essays on growth options and corporate strategy. Doctoral dissertation. The Ohio State University, Columbus, OH

Tong T, Li J (2008) Real options and MNE strategies in Asia Pacific. Asia Pac J Manage 25(1): 153–169

Tong TW, Reuer J, Building MC (2003) Switching options and coordination costs in multinational firms. Proceedings of the 4th Igms Ciber Research Forum, Temple University, Philadelphia, PA

Tong TW, Li J (2008) Real options and MNE strategies in Asia Pacific. Asia Pac J Manage 25(1):153–169

Tong TW, Reuer JJ (2006) Firm and industry influences on the value of growth options. Strateg Organ 4(1):71

Tong T, Reuer J, Building MC (2004) Corporate investment decisions and the value of growth options. Paper presented at the Proceedings of the 2004 annual international conference real options, Montreal, Canada

Tong TW, Reuer JJ, Peng MW (2008) International joint ventures and the value of growth options. Acad Manage J (AMJ) 51(5):1014–1029

Tranfield D, Smith S (1998) The strategic regeneration of manufacturing by changing routines. Work 18(2):114–129

Trippner K (2006) Systematische Risikobewertung in versorgungslogistischen Systemen in der Automobilindustrie: Analyse internationaler Hersteller in China. Universitätsbibliothek der BTU Cottbus

Tukiainen T (2004) The unexpected benefits of internal corporate ventures: an empirical examination of the consequences of investment in corporate ventures

Tushman M, Murmann JP (2002) Dominant designs, technology cycles, and organizational outcomes. Managing in the modular age: architectures, networks, and organizations

Tyler BB, Steensma HK (2007) Evaluating technological collaborative opportunities: a cognitive modeling perspective. Strateg Manage J 16(S1):43–70

Tyler BB, Steensma HK (1995) Evaluating technological collaborative opportunities: a cognitive modeling perspective. Strateg Manage J 43–70

Upson JW (2008) Why do firms exercise foothold options: explanations from multipoint competition and real options theory, Doctoral Dissertation. Florida State University, Tallahassee, FL

Van Bekkum S, Pennings E, Smit H (2009) A real options perspective on R&D portfolio diversification. Res Policy 38(7):1150–1158

Van de Vrande V, Lemmens C, Vanhaverbeke W (2006) Choosing governance modes for external technology sourcing. R&D Manage 36(3):347–363

van de Vrande V, Vanhaverbeke W, Duysters G (2009) External technology sourcing: the effect of uncertainty on governance mode choice. J Bus Ventur 24(1):62–80

Vanhaverbeke W, Van de Vrande V, Cloodt M, Agoralaan-building D, Gebouw D (2008) Connecting absorptive capacity and open innovation, Available at SSRN 1091265

Vanhaverbeke W, Duysters G (1997) A longitudinal analysis of the choice between technology-based strategic alliances and acquisitions in high-tech industries: the case of the ASIC industry. Paper presented at the innovation in technology management-the key to global leadership. PICMET'97: Portland international conference on management and technology, Portland, OR

Vanhaverbeke W, Duysters G, Noorderhaven N (2002) External technology sourcing through alliances or acquisitions: an analysis of the application-specific integrated circuits industry. Organ Sci 13(6):714–733

Vanhaverbeke W, Van de Vrande V, Chesbrough H, Field S, Agoralaan-building D, Gebouw D (2008) Understanding the advantages of open innovation practices in corporate venturing in terms of real options. Creat Innov Manage 17(4):251–258

Vaquero IE, Cruz NM, de la Fuente G. Technological strategic alliances: formation under the real options approach. Documento de Trabajo 4, 08

Vassolo RS, Anand J, Folta TB (2004) Non-additivity in portfolios of exploration activities: a real options-based analysis of equity alliances in biotechnology. Strateg Manage J 25(11): 1045–1061

Verbeeten FHM (2006) Do organizations adopt sophisticated capital budgeting practices to deal with uncertainty in the investment decision?: a research note. Manage Account Res 17(1): 106–120

Villaverde PMG (2007) M. ª J. Ruiz Ortega. Cuadernos de Economía y Direcci6n de la Empresa, 32: 111–148

Villaverde G, Manuel P, Ortega R, José M. Configuraciones Organizativas en Sectores Dinámicos y Hostiles: Adecuaci6n al Contexto Sectorial, Coherencia Interna y Resultados 3. Cuadernos de Economía y Direcci6n de la Empresa

Volberda HW, Elfring T (2001) Rethinking strategy. Sage Publications. Thousand Oaks, CA

Volume B. Coff RW, Laverty KJ (2007) Real options meet organizational theory: coping with path dependencies, agency costs, and organizational form. Advances in Strategic Management 24:333–361

von Diplom-Betriebswirt Andreas MK, Meyer KE, Bath G, Devinney TM, Sydney A. Market entry and expansion through international joint ventures: a multi-causal analysis of international joint venture performance

Von Kent Eriksson JJ, Majkgård A, Sharma DD. Time and experience in the internationalization process

Vonortas NS, Hertzfeld HR (1998) Research and development project selection in the public sector. J Policy Anal Manage 17(4):621–638

Wang SY, Lee CF. A fuzzy real option valuation approach to capital budgeting under uncertainty

Wang H, Lim SS (2008) Real options and real value: the role of employee incentives to make specific knowledge investments. Strateg Manage J 29(7):701–721

Warner AG, Fairbank JF, Steensma HK (2006) Managing uncertainty in a formal standards-based industry: a real options perspective on acquisition timing. J Manage 32(2):279

Weeds H (2006) Applying options games: when should real options valuation be used? Unpublished working paper. The University of Essex

West Iii GP, DeCastro J (2001) The achilles heel of firm strategy: resource weakness and distinctive inadequacies. J Manage Stud 38(3):417–442

Whittle JA, Captain U (2003) Human resource slack as an antecedent to instilling the entrepreneurial mindset within department of defense organizations: air force inst of tech Wright-Patterson afb oh school of engineering and management

Wiedenhofer M (2003) Bewertung von Kernkompetenzen: Strategische Ressourcen als Realoption: Deutscher Universitäts-Verlag

Wilk E. Programa de Pós-Graduação em Administração Mestrado em Administração

Wischnevsky JD, Damanpour F (2005) Punctuated equilibrium model of organizational transformation: Sources and consequences in the banking industry. Research in Organizational Change and Development, 15:207–239

Wischnevsky JD, Damanpour F (2005) Punctuated equilibrium model of organizational transformation: sources and consequences in the banking industry. Research in Organizational Change and Development, 207

Woiceshyn J, Daellenbach U (2005) Integrative capability and technology adoption: evidence from oil firms. Ind Corp Change 14(2):307–342

Wolfe R, Wright PM, Smart DL (2006) Response to editorial by Bryan and Rafferty. Hum Resour Manage 45(4):673–676

Xu D, Zhou C, Phan PH (2009) A real options perspective on sequential acquisitions in China. J Int Bus Stud 41(1):166–174

Yang H, Dess GG (2007) Where do entrepreneurial orientations come from? An investigation on their social origin. In: Entrepreneurial Strategic Processes. p 223

Yigang P, Chi PSK. Financial performance and survival of multinational corporations in China. Strateg Manage J 20(4): 359

Zardkoohi A (2004) Do real options lead to escalation of commitment? Acad Manage Rev 29(1): 111–119

Zubac A (2009) Investing in resources to create customer value: the organisational, strategic and performance implications, Doctoral Dissertation. University of Adelaide, Business School, Adelaide, Australia

Annex II
List of Survey Questions

The questions will allow us to test for the following hypotheses:

H_1: The use of real options in decision making for technology investments increases in a co-opetitive environment. (comparing subpopulations)

H_2: The use of real options in decision making for *operational, tactical, and strategic* technology investments increases with risk and uncertainty.

- H_{2A}: The use of real options in decision making for *strategic* technology investments increases with risk and uncertainty. (comparing subpopulations)
- H_{2B}: The use of real options in decision making for *nonstrategic* technology investments increases with risk and uncertainty. (comparing subpopulations)
- H_{2C}: The use of real options in decision making applies *more often for strategic than nonstrategic* technology investments. (in all subpopulations)

H_3: The use of real options in decision making increases when the technology investment under consideration leads to new business areas (product, services, and/or processes). (comparing subpopulations)

In order not to set specific expectations and bias the respondent, we will not refer directly to real options but to flexibility in investment (Busby and Pitts 1997).

Business Profile

Q.1. Your business is inclined toward: (low risk/high risk)[1] (filter)

- ☐ Low-risk projects with certain and normal rates of return
- ☐ A balance of low-risk and high-risk projects
- ☐ High-risk projects with chances of very high returns
- ☐ DK/refuse

[1] Question used in or adapted from an NFIB Small Business Poll.

C.M. Sipp and E.G. Carayannis, *Real Options and Strategic Technology Venturing: A New Paradigm in Decision Making*, SpringerBriefs in Business 31,
DOI 10.1007/978-1-4614-5814-2, © Springer Science+Business Media New York 2013

Q.2. Would it be most accurate to describe your business as:[3] (filter)
- ☐ Producer of technology
- ☐ Extensive user of technology
- ☐ Producer and user of technology
- ☐ Avoids technology
- ☐ Others, please specify
- ☐ DK/refuse

Q.3. Do you have a business website?[3] (filter)
- ☐ Yes
- ☐ No
- ☐ DK/refuse

Q.4. Is your Web site capable of conducting e-commerce, that is, accepting orders over the Internet and letting customers pay with a credit card through a secure Web connection?[3] (filter)
- ☐ Yes
- ☐ No
- ☐ DK/refuse

Q.5. When you consider making a technology investment in your business, how do you assess its financial viability? Do you primarily use (select all that apply): [2,3] (existence/nonexistence)
- ☐ Payback period
- ☐ Return on investment (ROI)
- ☐ Discounted cash flow methods (NVP or IRR)
- ☐ Gut feeling
- ☐ Others, please specify
- ☐ Not applicable – no major investments
- ☐ Don't know/refuse

Co-opetition

Please rate the importance of the following flexibility attributes influencing technology investments *in an area subject to both collaboration and competition with business partners*. Please consider investments made today, as well as in 2008 (precrisis). (H_1)

Today	Very important				Not important at all		2008	Very important				Not important at all	
	1	2	3	4	5	N/A		1	2	3	4	5	N/A
Q.6. Postponement	☐	☐	☐	☐	☐	☐	Postpone	☐	☐	☐	☐	☐	☐
Q.7. Abandonment	☐	☐	☐	☐	☐	☐	Abandon	☐	☐	☐	☐	☐	☐
Q.8. Rescale	☐	☐	☐	☐	☐	☐	Rescale	☐	☐	☐	☐	☐	☐
Q.9. Growth	☐	☐	☐	☐	☐	☐	Growth	☐	☐	☐	☐	☐	☐

[2] Question used in or adapted from Busby and Pitts (1997).

Coevolution

Please rate the importance of the following flexibility attributes influencing a *strategic technology investment* under risk and uncertainty. Please consider investments made today, as well as in 2008 (pre-crisis). (H_{2A}, H_{2D})

Strategic decision investment: leads to a substantial change in the business model and value proposition, including a shift into a different or emerging sector or business niche.

Today	Very important				Not important at all		2008	Very important				Not important at all	
	1	2	3	4	5	N/A		1	2	3	4	5	N/A
Q.10. Postponement	☐	☐	☐	☐	☐	☐	Postpone	☐	☐	☐	☐	☐	☐
Q.11. Abandonment	☐	☐	☐	☐	☐	☐	Abandon	☐	☐	☐	☐	☐	☐
Q.12. Rescale	☐	☐	☐	☐	☐	☐	Rescale	☐	☐	☐	☐	☐	☐
Q.13. Growth	☐	☐	☐	☐	☐	☐	Growth	☐	☐	☐	☐	☐	☐

Please rate the importance of the following flexibility attributes influencing a *nonstrategic technology investment* under risk and uncertainty. Please consider investments made today, as well as in 2008 (pre-crisis). (H_{2C}, H_{2D})

Today	Very important				Not important at all		2008	Very important				Not important at all	
	1	2	3	4	5	N/A		1	2	3	4	5	N/A
Q.14. Postponement	☐	☐	☐	☐	☐	☐	Postpone	☐	☐	☐	☐	☐	☐
Q.15. Abandonment	☐	☐	☐	☐	☐	☐	Abandon	☐	☐	☐	☐	☐	☐
Q.16. Rescale	☐	☐	☐	☐	☐	☐	Rescale	☐	☐	☐	☐	☐	☐
Q.17. Growth	☐	☐	☐	☐	☐	☐	Growth	☐	☐	☐	☐	☐	☐

Co-specialization

Please rate the importance of the various flexibility attributes influencing a technology investment *leading to new business areas* (products, services, processes). Please consider investments made today, as well as in 2008 (precrisis). (H_3)

Today	Very important				Not important at all		2008	Very important				Not important at all	
	1	2	3	4	5	N/A		1	2	3	4	5	N/A
Q.18. Postponement	☐	☐	☐	☐	☐	☐	Postpone	☐	☐	☐	☐	☐	☐
Q.19. Abandonment	☐	☐	☐	☐	☐	☐	Abandon	☐	☐	☐	☐	☐	☐
Q.20. Rescale	☐	☐	☐	☐	☐	☐	Rescale	☐	☐	☐	☐	☐	☐
Q.21. Growth	☐	☐	☐	☐	☐	☐	Growth	☐	☐	☐	☐	☐	☐

Respondent Profile

Q.22. What is your job title? ______________________________________

Q.23. How many people does your company employ?

- ☐ Less than 50
- ☐ 50–100
- ☐ 101–200
- ☐ 201–500
- ☐ 501–1,000
- ☐ 1,001–5,000
- ☐ More than 5,000
- ☐ Don't know/refuse to answer

Q.24. In which industry/sector does your company operate? (list of industries)________________________________

Q.25. In which country is your company headquartered? (list of countries)____________________________________

Q.26. If located in the USA, please specify in which state: (list of U.S. states)___
____

Q.27. Would you be available for a follow-up interview? If so, please provide your name and email and/or phone number:

____

Annex III
List of Interview Topics

Q1. Please describe the decision-making process of your company for technological investments. Please specify any differences there may be for a strategic versus non-strategic investments.

Q2. What factors are taken into account when deciding whether to invest in technology? Please specify which factors are explicitly taken into account (e.g., described in procedures) and which are implicit.

Q3. Suppliers
Q3a. In which industries/subsectors do your suppliers operate?

Q3b. Where are your suppliers located? (Select all that apply.)
__ Domestic
    __ National
    __ Regional
    __ Local
__ International

Q3c. How and to what extent do your suppliers influence your decision making in technology investments?

Q4. Buyers
Q4a. In which industries/subsectors do your buyers operate?

Q4b. Where are your buyers located? (Select all that apply.)
__ Domestic
    __ National
    __ Regional
    __ Local
__ International

Q4c. How and to what extent do your buyers influence your decision making in technology investments?

C.M. Sipp and E.G. Carayannis, *Real Options and Strategic Technology Venturing: A New Paradigm in Decision Making*, SpringerBriefs in Business 31,
DOI 10.1007/978-1-4614-5814-2, © Springer Science+Business Media New York 2013

Q5. Competitors

Q5a. In which industries/subsectors do your competitors operate?

Q5b. Where are your competitors located? (Select all that apply.)

__ Domestic

  __ National

  __ Regional

  __ Local

__ International

Q5c. How and to what extent do your competitors influence your decision making in technology investments?

Q6. Regulators

Q6a. If applicable, in which industries/subsectors do your regulators operate?

Q6b. Where are your regulators located? (Select all that apply.)

__ Domestic

  __ National

  __ Regional

  __ Local

__ International

Q6c. How and to what extent do your regulators influence your decision making in technology investments?

Annex IV
Simulation Design

This annex presents the simulation's design as described in Provance's (2010) dissertation and has been adapted to reflect some of the changes applied to this research.

> The simulation is designed as a [16×16] torus-shaped landscape (or lattice) to eliminate ordinal bias and bias against agents along edges of the landscape10. The contents of this landscape are knowledge resources that are acquired by new ventures and incumbent firms in a competition to fill market needs (Fig. A.1).

Figure A.1 depicts a portion of the landscape on which this simulation runs along with the heterogeneous set of agents that exist on that landscape. This simulation model includes incumbent firms, new ventures, buyers, institutions, and informal knowledge clusters in each "neighborhood" (or lattice point) on the landscape. The two former agents (firms and ventures) develop formal network connections with partners (other firms or ventures) during the execution of the simulation. Firms and ventures acquire informal knowledge clusters during the simulation. Institutions act more holistically on neighborhoods by mediating the flows of knowledge between other agents on the landscape.

Prior to initialization for a specific run, the simulation model dictates certain behaviors at a system level, shaping the landscape in order to reflect a desired set of conditions. Certain simulation parameters are set at a global level:

- The duration in time periods of an iteration of the simulation
- The number of iterations to run a particular configuration of the simulation for
- The seed key for random number generation during each iteration
- Initial levels of buyers, incumbent firms, incumbent firm network partners, and informal knowledge clusters
- Maximum distance for partner selection when growing networks
- Initialization conditions for incumbent firm, buyer, and informal cluster knowledge
- Rates of entry for new informal knowledge clusters, buyers, and new venture foundings
- The threshold for determining new venture creation (founder commitment to create an entrepreneurial venture)

C.M. Sipp and E.G. Carayannis, *Real Options and Strategic Technology Venturing: A New Paradigm in Decision Making*, SpringerBriefs in Business 31,
DOI 10.1007/978-1-4614-5814-2, © Springer Science+Business Media New York 2013

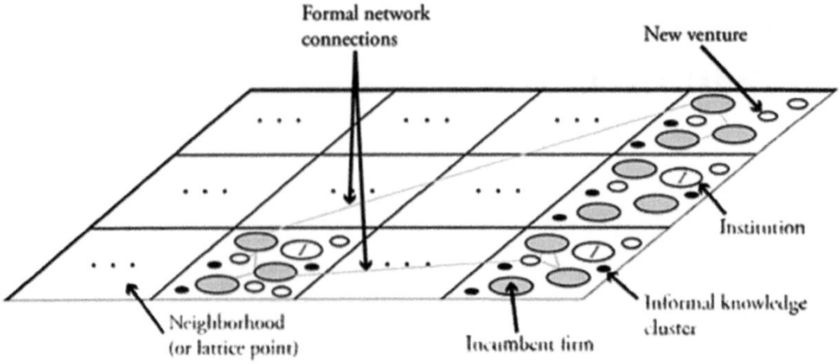

Fig. A.1 Conceptual model of heterogeneous agent landscape of simulation

- The threshold for entry by the new venture into the market competition (conversion of new venture into incumbent firm agent behaviors)
- The threshold for acceptance of innovation by market recipients
- Feedback thresholds and rates for observations of new venture failure and success
- Rates of change for early-adopter buyers

Other system parameters are configured at a lattice point level:

- Probabilities of new ventures and incumbent firms using informal and formal knowledge acquisition mechanisms
- Probabilities of using knowledge transformation occurring through serendipity or knowledge arbitrage
- The presence and type of institution mediating knowledge, configured as a weighting in the simulation.

At system initialization, each lattice point is populated with market recipients, incumbent firms, institutions, and loose knowledge clusters up to the levels specified by the configuration of the simulation run. Each of these agents is endowed with knowledge resource characteristics, as described below.

Simulation Agents

The primary agents in these systems include entrepreneurial founders, new ventures, incumbent firms ("competitors"), institutions, and market recipients of innovations ("buyers"). The underlying fundamental mechanism of behaviors and interactions of these agents is knowledge acquisition. The bases for the simulation are agent behaviors that dictate external knowledge acquisition events. The firm's external knowledge accumulation process is defined in the literature as the combination of acquisition and assimilation activities, although it goes by different labeling across papers. Todorova and Durisin (2007) summarize these differences (i.e., Cohen and Levinthal 1990;

Zahra and George 2002) in their reconceptualization of absorptive capacity. This study examines two significant components of this process: acquisition of knowledge from external sources, and transformation of acquired knowledge into useful internalized knowledge. In this simulation, the new venture is assumed to possess relatively little extant knowledge, so it must focus on acquisition from its inception. For this reason, other mechanisms of the knowledge appropriation process are held constant across agents, assuming each agent uses the knowledge acquired with perfect efficiency given constraints it has on market information and innovation capability (specified as the difference between its extant knowledge and the market need it resolves).

Knowledge possessed by an agent in this model is represented as a 28-bit binary string in which each bit represents distinct, specialized knowledge. An example of the string is depicted thusly: 0011 0101 1111 0111 1011 1100 0001. The same sized string is used for new ventures, incumbent firms, buyers, and informal knowledge clusters. The content of string differs at initialization in the numbers of bits that are "flipped" or changed from 0 to 1 during configuration. The string represents an innovation being developed by incumbent firms and new venture, an innovation "needed" by buyers, or a collection of knowledge that could be useful to an innovation (informal knowledge cluster), respectively. For example, a buyer string could reflect requirements for a new product that is needed, while the corresponding string of a new venture may be the innovation that fills that need.

Firms

Firms are incumbent competitors introduced at the beginning of a simulation run. New ventures that meet an entry threshold described below may become "firms," too. For the purposes of defining simulation configuration initialized for each run, only the first case is addressed here. A level of six firms is configured in each lattice point at the outset of a simulation run. Each firm receives a knowledge endowment of randomly assigned 1 s in 7–43 % of the available positions in its knowledge string, as shown here:

Incumbent firm configuration at initialization:

0101101011110000000111000001

Buyers and Needs

Recipients of innovations, or "buyers," possess market needs. The specifications of these needs are not unique products, but rather a combination of requirements and preferences within a product-market domain that exhibits rarity. Different configurations of market needs exist within distinct opinion clusters in a market (Provance and DiGregorio, 2007). The needs of market recipients are expressed in the simulation as a market need string that is similar to the knowledge resource string of firms. The market need string represents a target for innovation activities of firms and new ventures. Where knowledge resource strings are composed of specialized capabilities and resources in the firm or venture, market need strings can be thought of as combinations of requirements and preferences. That is, a product-market domain may have several designs or perhaps may evolve over time to a dominant design.

The number of market needs is established at a lattice point level. The simulation used for this study distributes market needs at equal levels (6 per lattice point) across all lattice points. The simulation adds new market needs according to a probability distributed seeded with a rate configured during initialization. Market needs are also removed when the need has been satisfied by an innovation supplied from an incumbent firm or new venture. Each market need is configured initially with a randomly assigned endowment of 1 s in 50 % or greater of the available positions in the market need string. For example, an initial configuration might look like the following:

Market need configuration at system initialization:

0111110110101111011010011011

Informal Knowledge Clusters (Ideas)

The ideas on the landscape represent the possibilities for informal acquisition of new knowledge. Ideas are accessed through social interaction of market participants, such as that described by Saxenian (1994) and others regarding the emergence and growth of Silicon Valley. Engineers would converse in social settings, which revealed interesting problems and identified new approaches to solutions that would lead to innovations. These interactions occurred not because of the formal relations the engineers' firms had established. Rather, proximity produced interactions. Informal access to knowledge in the simulation is modeled as the availability of loose knowledge clusters that new ventures and firms can acquire, which lead to new ideas. The process of combination is discussed below in the "Knowledge Acquisition" section.

The number of informal knowledge clusters is established at a lattice point level. In this simulation model, informal knowledge clusters are distributed equally across all lattice points and were initially set at six informal knowledge clusters per lattice point for this study. These clusters grow at a rate prescribed during configuration and are removed when used by new ventures or incumbent firms. Informal knowledge clusters are created at initialization by randomly assigning three (3) to four (4) bits of the informal knowledge cluster string as 1s. Any of the 28 bits could be assigned a setting of 1 with equal probability. For example, an initial configuration might look like the following:

Idea configuration at system initialization:

0100001000000000000100000000

Potential New Ventures and New Ventures

A simulation run begins with only incumbent firms, ideas, and market recipients existing on the landscape. Once the run starts, the possibility for the creation of new ventures exists. This process occurs in two stages: founder commitment and new venture formation. Founder commitment involves the decision making of a nascent entrepreneur that leads to the creation of a new venture. Potential new ventures are created according to a probability distribution based on seed rate defined at initialization. Potential venture knowledge endowments are granted to generated founders by assigning 1s to zero (0) to four (4) bits of the knowledge string on a random basis with equal probability. An example of an initial configuration of the potential venture knowledge endowment (PVKE) is illustrated below.

$$\text{PVKE endowment at creation}: 0100\,0011\,0000\,0000\,0001\,0000\,0000$$

Simulation System Events

In modeling the new venture formation process, the creation of a new venture (i.e., incorporating the business, opportunity recognition) is separated from formation or the early-stage efforts to develop a product/service and enter the market (e.g., pre-competitive R + D). New ventures exist in the simulation in one of four states: potential new venture, forming new venture, formed (or survived) new venture, and failed new venture. New ventures move through these states based on their activities in events related to knowledge acquisition, transformation, and use. The simulation model consists of seven events that run on a scheduled basis over the duration of the simulation:

(0) Founder commitment and new venture creation
(1) Capability generation
(2) New venture transition to incumbent firm
(3) Incumbent firm product innovation
(4) Product innovation adoption
(5) Buyer resolution → replacement of filled need
(6) Institutional knowledge mediation

All seven events are scheduled in every lattice point at rates determined by an exponential probability distribution. These rates are further affected by feedback loops that speed up or slow down formation activities based on prior failures, successes, and other observable recent history in the lattice point. Every live agent in the system is active in one of these event states, based on their rate of action and characteristics. Each agent exhibits behaviors appropriate to that state when scheduled to undertake it. The system of events is depicted in Fig. A.2.

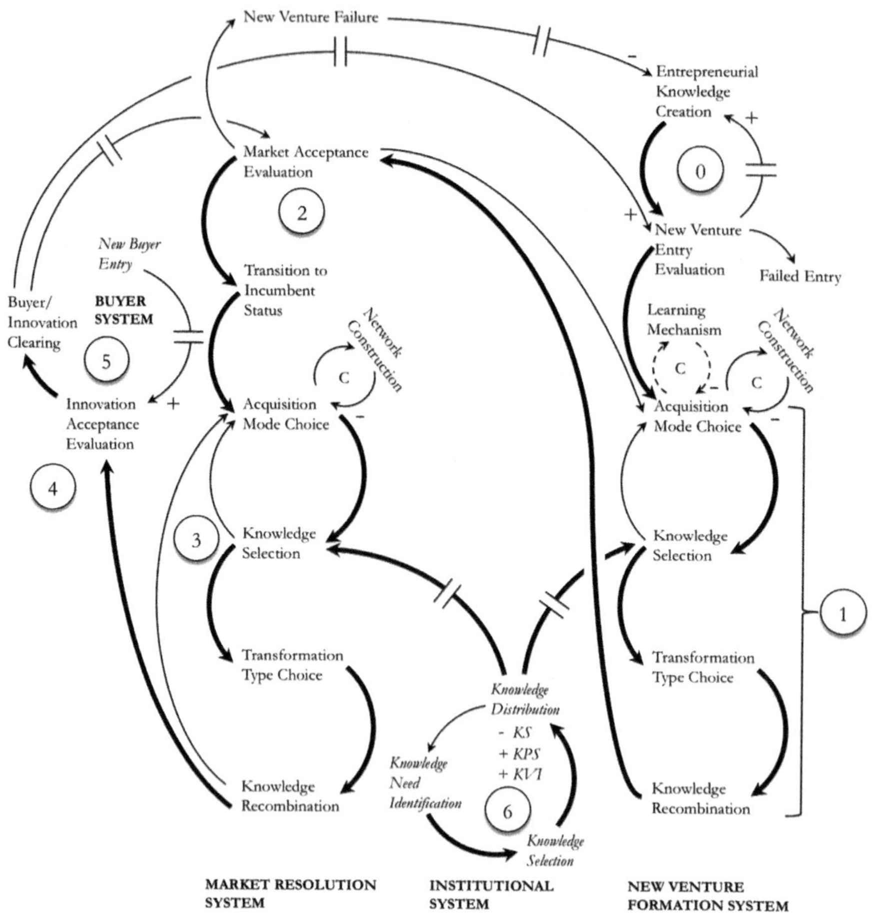

Fig. A.2 System model of new venture formation simulation

Founder Commitment (0)

This event refers to the entrepreneur's decision to pursue a new venture. That is, it represents the move from nascent or potential entrepreneurship to realized entrepreneurial activity. In the simulation, we assume that the decision to create is related to recognition of a need on the landscape. This recognition is reflected by the degree to which a founder's vision matches a particular market need in the same lattice point.

In each period, the PVKE generated is compared to market need strings in the same lattice point. If it matches (meaning 1s in each string are aligned) greater than 7% of the overall string comparison, then a new venture is created, and it enters stage 1, the new venture formation subsystem where knowledge acquisition occurs. This threshold is based on tuning the simulation to rates of realized entrepreneurship (from nascent entrepreneurship) established by other data sources (University of Michigan, 2009).

New Venture Formation (1)

This event addresses the growth processes associated with new ventures following their founding. In this way, it is distinct from the act of creation, which has been linked to initial discovery or other value recognition actions (Alvarez and Barney 2005; Kirzner 2008; Sarasvathy 2008). Specifically, new venture formation is conceived in this model as a capability generation process based on the external acquisition of knowledge. The mode of an entrepreneur's knowledge acquisition process is manipulated between informal and formal mechanisms to represent one variable of interest in this study.

This event is operationalized as a set of decisions made by the entrepreneur regarding knowledge acquisition and transformation choices. In this study, knowledge transformation is held constant as a knowledge arbitrage function, which is the purposeful recombination of acquired knowledge with existing knowledge based on the unique perception and recognition of value by the entrepreneur (Carayannis 2008). When this event occurs, the rule the new firm agent follows is to acquire new information in an attempt to improve its position relative to a market need in the same lattice point when it is transformed into internalized knowledge. "Improving a position" means coming closer to a match with the market need. This rule is modified by another rule that introduces variation and suboptimal behaviors. This rule uses an entropy algorithm to reflect experience in strategic decision making (Neyman and Okada 2000). New ventures generally lack the same level of organizational decision-making experience as incumbents (Audretsch and Lehmann 2005), which may lead to deficiencies in performance. In the simulation, the effects of entropy are reversed in order to simulate experience. The new venture begins its formation in a "hot" condition of entropy. The hotter the condition, the more likely the new venture is to select detrimental knowledge leading to a suboptimal position. As time passes and knowledge is acquired, the venture's entropic status "cools," which increases the likelihood that the growing venture will choose knowledge that improves its position.

Knowledge Acquisition. Variation in the knowledge acquisition process is introduced into the system based on the formality of interaction between source and recipient. Specifically, two levels of formality are used: informal and formal. Informal knowledge acquisition occurs through the incidental interaction of agents and is biased toward more localized interaction. An example from Saxenian (1994) is two engineers from technology companies in Silicon Valley meeting at a bar. In the course of discussion, they exchange ideas that lead one engineer to solve a challenging technological issue with a new product. Formal knowledge acquisition occurs at the firm level through a network of obligated relationships. These relationships may range from transactional to strategic but are formalized through contracts or other mechanisms. Supplier-customer relationships and strategic alliances are examples of this type of acquisition.

In the simulation, informal knowledge acquisition is modeled as the new venture using the knowledge contained in the string of an informal knowledge cluster that exists within the same lattice point. Formal knowledge acquisition is modeled as knowledge procured by the firm from information flowing to the firm from a partner in the firm's network of relationships. This network has a broader distribution on the landscape than the distribution of available informal knowledge clusters in order to reflect the more geographically dispersed nature of alliance networks (Almeida and Kogut 1997; Rosenkopf and Almeida 2003). If a new venture is configured to prefer to acquire knowledge formally, then it will select a partner from its formal knowledge network or create a relationship with a new partner not already in its network. If the new venture does not possess any relationships, then it will create a new one as a default in order to acquire knowledge formally. The probability distribution of friends attaching to a firm's network is an inverse exponential function based on concentric rings emanating away for the firm at the focal point. The first ring has a probability of 0.5, the second ring a probability of 0.25, the third ring a probability of 0.125, and so on until the distance established during configuration has been covered.

The firm will evaluate the three (3) bit sets of knowledge from the partner's knowledge resource string until it identifies a set that improves its position relative to a buyer in its lattice point. This selection represents a local optimum for knowledge acquisition. Also, the effects of entropic behavior may hamper the selection when the new venture is younger, as was described above. It will select the three bits randomly with equal probability across all bits in the knowledge resource string. Acquisition is unidirectional. There is no quid pro quo exchange of knowledge, although the partner could select the original firm for acquisition during its turn. The model is configured with a probability (0.0–1.0) of selecting formal knowledge acquisition at initialization. This probability is established independently for each lattice point based on the design of a particular simulation run. In both types, the choice about which information to acquire from an informal knowledge cluster or node in formal network is based on identifying an improvement in the new venture's resource string relative to one of the market needs in the same lattice point (matching process). This behavioral rule is modified by an entropy-based choice rule that may reduce the improvement or even cause the new firm to choose knowledge that puts it into a lower position relative to market needs.

Knowledge Transformation. New ventures acquire new knowledge through formal network connections (e.g., alliances) or by absorbing localized spillovers of knowledge. Once acquired, the knowledge is transformed into innovation capabilities by combining it with the agent's existing knowledge resources. In the simulation, the knowledge transformation function recombines knowledge resources possessed by the new venture with external knowledge acquired formally or informally. This study treats knowledge transformation as either serendipity or knowledge arbitrage, based on the configuration being run. The mode of knowledge transformation is determined probabilistically during this event based on the serendipity parameter set for the respective lattice point. This decision is made in parallel with acquisition in order to determine how the knowledge possibilities are evaluated.

Under serendipity, knowledge transformation is the unexpected addition of valuable knowledge without discarding existing knowledge. The transformation of the knowledge under conditions of serendipity would follow an additive methodology (0101 0000 0000…+ 0110 1000 0000… becomes 0111 1000 0000…) in order to reflect the concept of fortuitous discovery and recognition. Under knowledge arbitrage conditions, the behavior exhibited by the new venture agent under this formation mechanism would be described as purposive action, actively seeking knowledge elements, or clusters that improve the market fit of its innovation (Carayannis, 2009). The transformation of the knowledge would follow a replacement methodology (0101 0000 0000…+ 0110 1000 0000…becomes 0110 1000 0000…) in order to reflect the concept of reallocation and recombination expressed in the original definition. That is, the venture forms as the combination of two existing knowledge configurations (the venture's extant one from the PVKE and the externally acquired information).

New Venture Market Entry (2)

New venture market entry is the matching process for converting a new venture into an incumbent firm. Essentially, it is the threshold at which the risk of failure diminishes substantially for a new venture because it has accumulated sufficient resources to weather adverse conditions or actions. In the simulation, this event is modeled as how well a new venture's approach (configuration of knowledge resources) is accepted by a market need in the same lattice point. If a new venture's knowledge resource string matches with one market need in the same lattice point by 60% (current default for venture threshold variable) or greater of the 1s in the bit string, it exits the new venture formation stage and becomes an incumbent firm. This matching process is demonstrated in Table A.1, in which new venture knowledge resource string matches 60% of a buyer need configuration.[3]

Table A.1 Demonstration of knowledge matching method for thresholds

New venture knowledge resource string	0111	1001	0110	1001	0101	1011	1100
Buyer need configuration	0111	0111	0110	0111	0111	1010	1111
Matches	111	1	11	1	11	1	11

[3] 12 "flipped" or activated new venture knowledge bits align with activated bits in the buyer need string out of 20 possible locations (12/20 = 60%).

Incumbent Firm Knowledge Acquisition (3)

Incumbent firms compete for the same knowledge resources on the landscape as new ventures. They begin with advantages over the new ventures because they are configured to possess formal networks with other incumbent firms. An assumption of this model is that exploitation of the formal networks for information acquisition will substantially overwhelm any exploratory effects of informal acquisition in large organizations (March, 1991). Thus, in the simulation design incumbent firm agents are imbued with only the formal information acquisition mechanism.

Firm Innovation Market Acceptance (4)

This stage is a matching process that clears firms' innovations from their ongoing strategic actions once they meet a market need in the same lattice point. This simulated matching is conceptually equivalent with launching a new product within a specific market niche, and having it accepted in the market. In the simulation, we model the process as a matching one in which the firm's knowledge resource string is compared to market needs in the same lattice point. If the alignment of 1 s in the two strings is 80 % or greater, the firm's innovation is accepted.

Market Need Resolution (5)

This final stage represents the replacement of filled market needs with new ones. In this first version of the simulation, a constant level of market needs is maintained within a lattice point, once the level is established at the initialization of the run. At this stage in the simulation, the filled market need is logged for location (lattice point) and time (period) and then removed from run. The simulation does not allow for market needs to become unfilled in this version. The firm filling the need is also logged as having filled one.

Institutional Knowledge Mediation (6)

Institutional knowledge mediation is the primary mechanism of interest in this investigation. Three distinct types of institutions of developed and employed in the simulation: (1) knowledge-standardizing, (2) knowledge production-stimulating, and (3) knowledge variation-inducing institutions. A central thesis of the study results presented in Provance (2010) is that new ventures interact differently with these types of institutions, which results in different levels of sustainable entrepreneurship regionally and performance at the new venture level of analysis. In the simulation, an institution is a function (rather than an agent) within a lattice point and acts based on the conditions of that lattice point at a given time. Its effects have the potential to spill over to surrounding lattice points based on the radius settings in the configuration, however.

Model Robustness and Stability Testing

As any model, this simulation proves useful if the results generated are proven to be a reflection of the agents' behavior rather than "artifacts of the programming or simulation architecture" (Provance 2010). Provance tested "the new venture simulation model was tested for both robustness and stability (Bar-Yam 1997). This stress testing consisted of two steps: robustness of results based on landscape, and stability of results over time." The analysis showed that the model was robust and that stable performance was achieved around time period 50.

Summary of Simulation Parameters and Variables

Variable	Obs	Mean	Std. dev.	Min	Max	Description
Configuration 1						
Agentid	30,600	5,099.52	2,944.57	–	10,217.00	Agent ID number
Birth	30,600	17.06	13.95	–	49.99	Period of new venture creation
x	30,600	7.51	4.09	–	15.00	X position on lattice
y	30,600	7.43	4.60	–	15.00	Y position in lattice
Config	30,600	1.00	–	1.00	1.00	Configuration label
Run	30,600	1.00	0.82	–	2.00	Run number
Group	30,600	1.18	0.65	–	2.00	Region on lattice
Age	30,600	27.28	16.46	0.01	50.00	Average age of agent in lattice point
Agenv	25,992	23.25	14.54	0.01	50.00	Average age of new venture in lattice point
Ageform	13,868	5.61	3.50	0.02	21.48	Average age of forming new venture in lattice point
Nvfail	25,992	0.38	0.48	–	1.00	Average level of new venture failures
Nvform	25,992	0.09	0.29	–	1.00	Avg. level of new ventures still forming
Nvsurv	25,992	0.53	0.50	–	1.00	Average level of new ventures formed
Formal	30,374	13.93	8.88	1.00	39.00	Average number of formal acquisition actions
Netsize	30,374	4.68	2.79	1.00	15.00	Average network size of new venture

(continued)

(continued)

Variable	Obs	Mean	Std. dev.	Min	Max	Description
Netlev	30,374	1.32	0.47	1.00	2.00	Average closeness centrality of new venture
Netcent	30,600	0.60	0.26	–	1.00	Ratio of the number of connections to the number of agents in the network
Configuration 2						
Agentid	32,155	16,077.00	9,282.49	–	32,154.00	Agent ID number
Birth	32,155	16.97	13.71	–	49.99	Period of new venture creation
x	32,155	7.46	4.27	–	15.00	X position on lattice
y	32,155	7.54	4.60	–	15.00	Y position in lattice
Config	32,155	2.00	–	2.00	2.00	Configuration label
Run	32,155	1.00	0.81	–	2.00	Run number
Group	32,155	2.87	1.38	–	5.00	Region on lattice
Age	–					Average age of agent in lattice point
Agenv	27,547	23.27	14.52	0.01	50.00	Average age of new venture in lattice point
Ageform	14,471	5.75	3.47	0.03	19.39	Average age of forming new venture in lattice point
Nvfail	27,547	0.39	0.49	–	1.00	Average level of new venture failures
Nvform	27,547	0.08	0.28	–	1.00	Avg. level of new ventures still forming
Nvsurv	27,547	0.53	0.50	–	1.00	Average level of new ventures formed
Formal	31,975	13.78	8.84	1.00	39.00	Average number of formal acquisition actions
Netsize	31,975	1,196.66	2,283.31	1.00	16,428.00	Average network size of new venture
Netlev	31,975	1.34	0.47	1.00	2.00	Average closeness centrality of new venture
Netcent	32,155	0.60	0.26	–	1.00	Ratio of the number of connections to the number of agents in the network

Annex V
Survey TD and TB Profiles

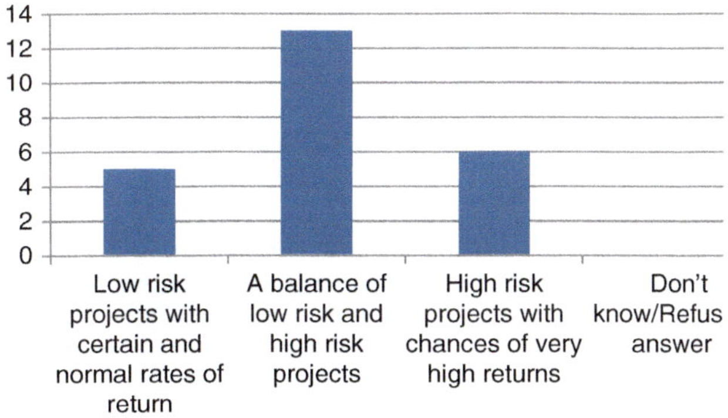

TD risk profile

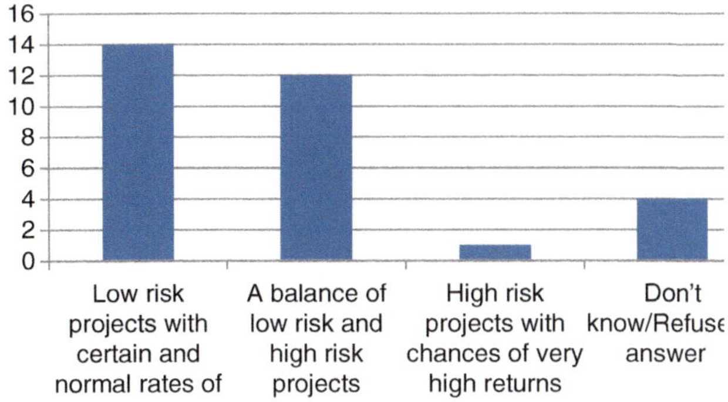

TB risk profile

C.M. Sipp and E.G. Carayannis, *Real Options and Strategic Technology Venturing: A New Paradigm in Decision Making*, SpringerBriefs in Business 31,
DOI 10.1007/978-1-4614-5814-2, © Springer Science+Business Media New York 2013

TD size profile

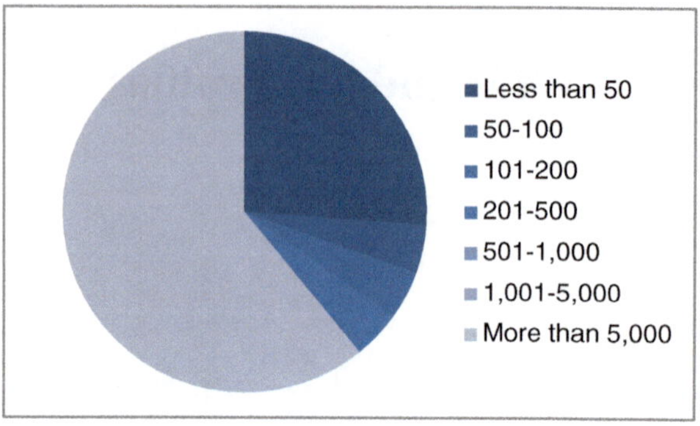

TB size profile

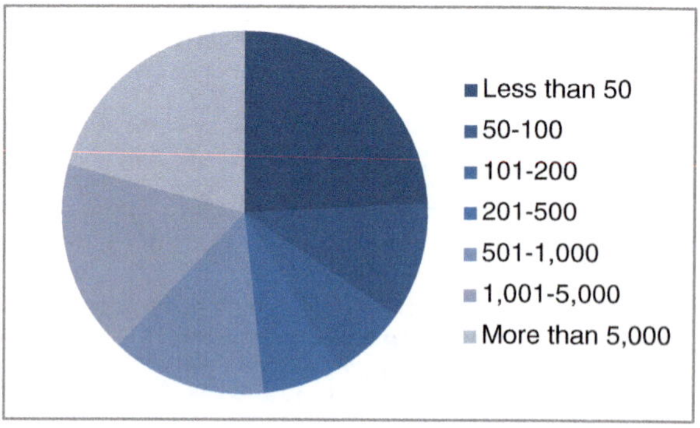

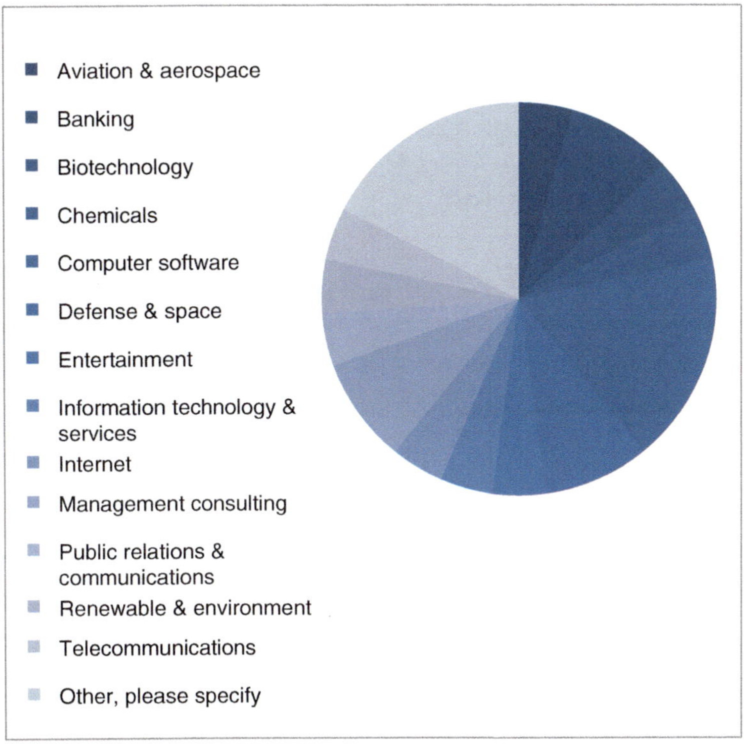

TD industry distribution

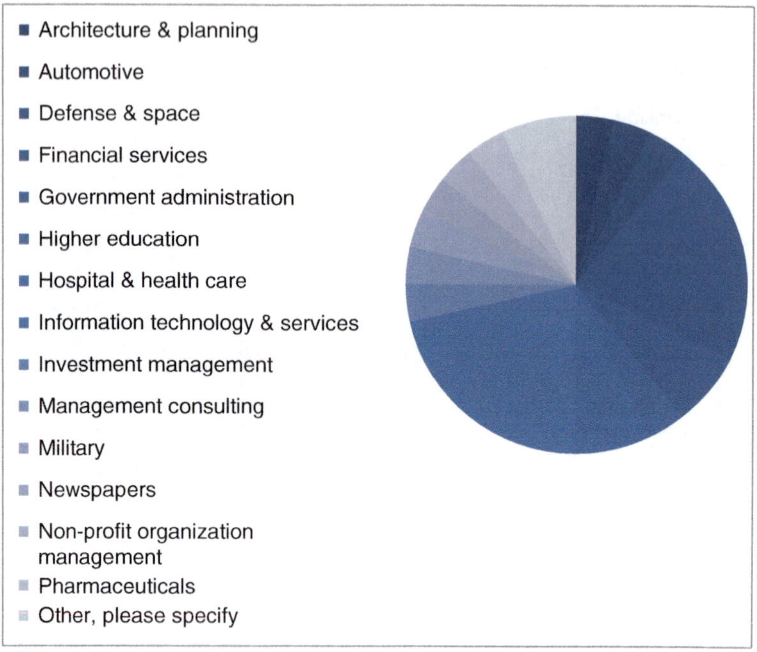

TB industry distribution

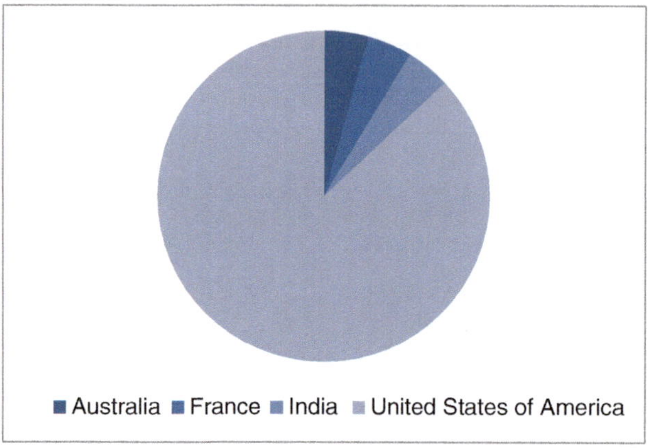

TD headquarters country distribution

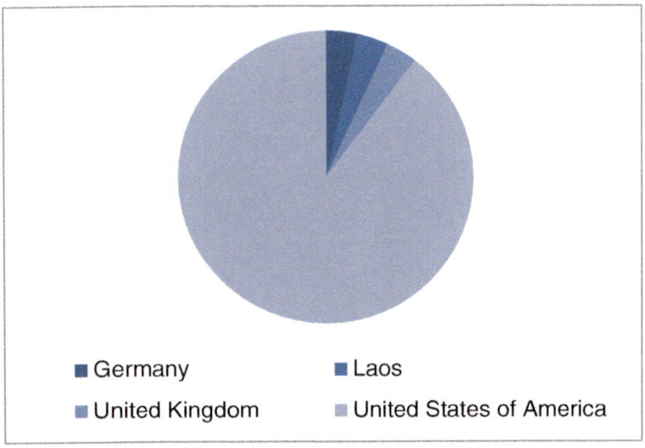

TD headquarters country distribution

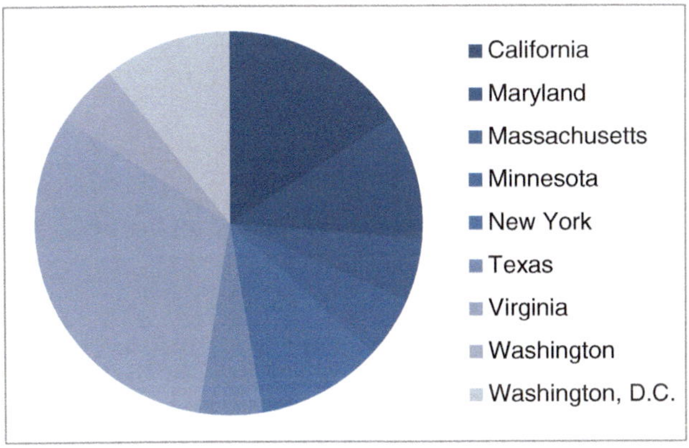

TD headquarters country distribution

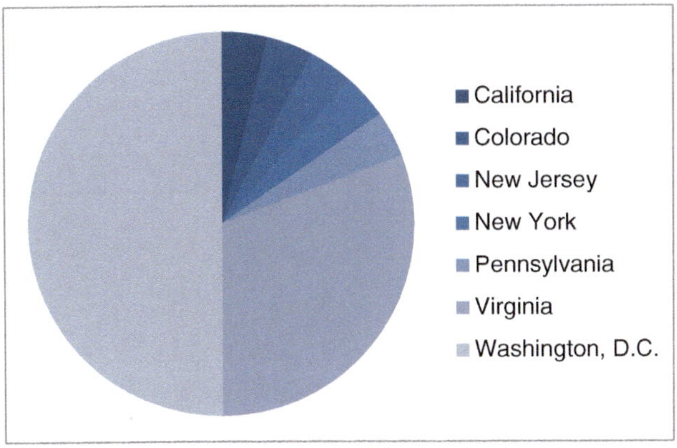

TD headquarters country distribution

Annex VI
Wilcoxon Rank-Sum (Mann–Whitney) Tests

```
3Cs Today&2008  Tuesday February 15 23:17:38 2011   Page 1

                                         STATA(R)
                                   Statistics/Data Analysis

                                      User: 3Cs Today&2008
                                   Project: Dissertation

1 . insheet using "C:\Users\caco\Desktop\Dissertation Process\DataCollection\Survey\Survey Res
(47 vars, 55 obs)

2 . do "C:\Users\caco\Desktop\Dissertation Process\DataCollection\Survey\StataAnalysis\3Cs Tod

3 . ranksum c1postpone, by(subpop)

Two-sample Wilcoxon rank-sum (Mann-Whitney) test

      subpop |      obs    rank sum    expected
-------------+---------------------------------
           2 |       30         898         825
           3 |       24         587         660
-------------+---------------------------------
    combined |       54        1485        1485

unadjusted variance     3300.00
adjustment for ties     -275.35
                      ----------
adjusted variance       3024.65

Ho: c1post~e(subpop==2) = c1post~e(subpop==3)
             z =   1.327
    Prob > |z| =   0.1844

4 . ranksum c1postpone2008, by(subpop)

Two-sample Wilcoxon rank-sum (Mann-Whitney) test

      subpop |      obs    rank sum    expected
-------------+---------------------------------
           2 |       30         851         825
           3 |       24         634         660
-------------+---------------------------------
    combined |       54        1485        1485

unadjusted variance     3300.00
adjustment for ties     -160.38
                      ----------
adjusted variance       3139.62

Ho: c1p~2008(subpop==2) = c1p~2008(subpop==3)
             z =   0.464
    Prob > |z| =   0.6426
```

C.M. Sipp and E.G. Carayannis, *Real Options and Strategic Technology Venturing: A New Paradigm in Decision Making*, SpringerBriefs in Business 31,
DOI 10.1007/978-1-4614-5814-2, © Springer Science+Business Media New York 2013

```
3Cs Today&2008   Tuesday February 15 23:17:39 2011   Page 2

5 . ranksum clabandon, by(subpop)

Two-sample Wilcoxon rank-sum (Mann-Whitney) test

      subpop |      obs    rank sum    expected
-------------+---------------------------------
           2 |       30       940.5         825
           3 |       24       544.5         660
-------------+---------------------------------
    combined |       54        1485        1485

unadjusted variance     3300.00
adjustment for ties     -180.75
                      ----------
adjusted variance       3119.25

Ho: claban~n(subpop==2) = claban~n(subpop==3)
             z =   2.068
    Prob > |z| =   0.0386

6 . ranksum clabandon2008, by(subpop)

Two-sample Wilcoxon rank-sum (Mann-Whitney) test

      subpop |      obs    rank sum    expected
-------------+---------------------------------
           2 |       30         859         825
           3 |       24         626         660
-------------+---------------------------------
    combined |       54        1485        1485

unadjusted variance     3300.00
adjustment for ties     -178.24
                      ----------
adjusted variance       3121.76

Ho: cla~2008(subpop==2) = cla~2008(subpop==3)
             z =   0.609
    Prob > |z| =   0.5428

7 . ranksum clrescale, by(subpop)

Two-sample Wilcoxon rank-sum (Mann-Whitney) test

      subpop |      obs    rank sum    expected
-------------+---------------------------------
           2 |       30       914.5         825
           3 |       24       570.5         660
-------------+---------------------------------
    combined |       54        1485        1485

unadjusted variance     3300.00
adjustment for ties     -243.40
                      ----------
adjusted variance       3056.60

Ho: clresc~e(subpop==2) = clresc~e(subpop==3)
             z =   1.619
    Prob > |z| =   0.1055
```

```
3Cs Today&2008   Tuesday February 15 23:17:39 2011   Page 3

 8 . ranksum c1rescale2008, by(subpop)

Two-sample Wilcoxon rank-sum (Mann-Whitney) test

      subpop |      obs    rank sum    expected
-------------+---------------------------------
           2 |       30       922.5         825
           3 |       24       562.5         660
-------------+---------------------------------
    combined |       54        1485        1485

unadjusted variance     3300.00
adjustment for ties     -210.82
                       ----------
adjusted variance       3089.18

Ho: c1r~2008(subpop==2) = c1r~2008(subpop==3)
             z =   1.754
    Prob > |z| =   0.0794

 9 . ranksum c1growth, by(subpop)

Two-sample Wilcoxon rank-sum (Mann-Whitney) test

      subpop |      obs    rank sum    expected
-------------+---------------------------------
           2 |       30        1002         825
           3 |       24         483         660
-------------+---------------------------------
    combined |       54        1485        1485

unadjusted variance     3300.00
adjustment for ties     -368.68
                       ----------
adjusted variance       2931.32

Ho: c1growth(subpop==2) = c1growth(subpop==3)
             z =   3.269
    Prob > |z| =   0.0011

10 . ranksum c1growth2008, by(subpop)

Two-sample Wilcoxon rank-sum (Mann-Whitney) test

      subpop |      obs    rank sum    expected
-------------+---------------------------------
           2 |       30       933.5         825
           3 |       24       551.5         660
-------------+---------------------------------
    combined |       54        1485        1485

unadjusted variance     3300.00
adjustment for ties     -256.86
                       ----------
adjusted variance       3043.14

Ho: c1g~2008(subpop==2) = c1g~2008(subpop==3)
             z =   1.967
    Prob > |z| =   0.0492
```

```
3Cs Today&2008   Tuesday February 15 23:17:39 2011    Page 4

11 . ranksum c1flex, by(subpop)

   Two-sample Wilcoxon rank-sum (Mann-Whitney) test

          subpop |      obs    rank sum    expected
   --------------+---------------------------------
               2 |       30         963         825
               3 |       24         522         660
   --------------+---------------------------------
        combined |       54        1485        1485

   unadjusted variance      3300.00
   adjustment for ties       -87.30
                          ----------
   adjusted variance        3212.70

   Ho: c1flex(subpop==2) = c1flex(subpop==3)
                z =   2.435
       Prob > |z| =   0.0149

12 . ranksum c1flex2008, by(subpop)

   Two-sample Wilcoxon rank-sum (Mann-Whitney) test

          subpop |      obs    rank sum    expected
   --------------+---------------------------------
               2 |       30         871         825
               3 |       24         614         660
   --------------+---------------------------------
        combined |       54        1485        1485

   unadjusted variance      3300.00
   adjustment for ties       -42.14
                          ----------
   adjusted variance        3257.86

   Ho: c1f~2008(subpop==2) = c1f~2008(subpop==3)
                z =   0.806
       Prob > |z| =   0.4203

13 . ranksum c2spostpone, by(subpop)

   Two-sample Wilcoxon rank-sum (Mann-Whitney) test

          subpop |      obs    rank sum    expected
   --------------+---------------------------------
               2 |       28       780.5         742
               3 |       24       597.5         636
   --------------+---------------------------------
        combined |       52        1378        1378

   unadjusted variance      2968.00
   adjustment for ties      -219.95
                          ----------
   adjusted variance        2748.05

   Ho: c2spos~e(subpop==2) = c2spos~e(subpop==3)
                z =   0.734
       Prob > |z| =   0.4627
```

```
3Cs Today&2008   Tuesday February 15 23:17:39 2011   Page 5

14 . ranksum c2spostpone2008, by(subpop)

Two-sample Wilcoxon rank-sum (Mann-Whitney) test

      subpop |      obs    rank sum    expected
-------------+---------------------------------
           2 |       28         779         742
           3 |       24         599         636
-------------+---------------------------------
    combined |       52        1378        1378

unadjusted variance     2968.00
adjustment for ties     -167.87
                      ----------
adjusted variance       2800.13

Ho: c2spos~8(subpop==2) = c2spos~8(subpop==3)
             z =   0.699
    Prob > |z| =   0.4844

15 . ranksum c2sabandon, by(subpop)

Two-sample Wilcoxon rank-sum (Mann-Whitney) test

      subpop |      obs    rank sum    expected
-------------+---------------------------------
           2 |       28         765         742
           3 |       24         613         636
-------------+---------------------------------
    combined |       52        1378        1378

unadjusted variance     2968.00
adjustment for ties     -205.25
                      ----------
adjusted variance       2762.75

Ho: c2saba~n(subpop==2) = c2saba~n(subpop==3)
             z =   0.438
    Prob > |z| =   0.6617

16 . ranksum c2sabandon2008, by(subpop)

Two-sample Wilcoxon rank-sum (Mann-Whitney) test

      subpop |      obs    rank sum    expected
-------------+---------------------------------
           2 |       28       750.5         742
           3 |       24       627.5         636
-------------+---------------------------------
    combined |       52        1378        1378

unadjusted variance     2968.00
adjustment for ties     -186.50
                      ----------
adjusted variance       2781.50

Ho: c2saba~8(subpop==2) = c2saba~8(subpop==3)
             z =   0.161
    Prob > |z| =   0.8720
```

```
3Cs Today&2008   Tuesday February 15 23:17:39 2011   Page 6

17 . ranksum c2srescale, by(subpop)

Two-sample Wilcoxon rank-sum (Mann-Whitney) test

      subpop |      obs    rank sum    expected
-------------+---------------------------------
           2 |       28         829         742
           3 |       24         549         636
-------------+---------------------------------
    combined |       52        1378        1378

unadjusted variance     2968.00
adjustment for ties     -269.99
                     ----------
adjusted variance       2698.01

Ho: c2sres~e(subpop==2) = c2sres~e(subpop==3)
             z =   1.675
    Prob > |z| =   0.0939

18 . ranksum c2srescale2008, by(subpop)

Two-sample Wilcoxon rank-sum (Mann-Whitney) test

      subpop |      obs    rank sum    expected
-------------+---------------------------------
           2 |       28         770         742
           3 |       24         608         636
-------------+---------------------------------
    combined |       52        1378        1378

unadjusted variance     2968.00
adjustment for ties     -263.28
                     ----------
adjusted variance       2704.72

Ho: c2sres~8(subpop==2) = c2sres~8(subpop==3)
             z =   0.538
    Prob > |z| =   0.5903

19 . ranksum c2sgrowth, by(subpop)

Two-sample Wilcoxon rank-sum (Mann-Whitney) test

      subpop |      obs    rank sum    expected
-------------+---------------------------------
           2 |       28         892         742
           3 |       24         486         636
-------------+---------------------------------
    combined |       52        1378        1378

unadjusted variance     2968.00
adjustment for ties     -343.22
                     ----------
adjusted variance       2624.78

Ho: c2sgro~h(subpop==2) = c2sgro~h(subpop==3)
             z =   2.928
    Prob > |z| =   0.0034
```

```
3Cs Today&2008   Tuesday February 15 23:17:39 2011   Page 7

20 . ranksum c2sgrowth2008, by(subpop)

Two-sample Wilcoxon rank-sum (Mann-Whitney) test

      subpop |      obs    rank sum    expected
-------------+---------------------------------
           2 |       28       884.5         742
           3 |       24       493.5         636
-------------+---------------------------------
    combined |       52        1378        1378

unadjusted variance      2968.00
adjustment for ties      -240.98
                       ----------
adjusted variance        2727.02

Ho: c2sgro~8(subpop==2) = c2sgro~8(subpop==3)
             z =   2.729
    Prob > |z| =   0.0064

21 . ranksum c2sflex, by(subpop)

Two-sample Wilcoxon rank-sum (Mann-Whitney) test

      subpop |      obs    rank sum    expected
-------------+---------------------------------
           2 |       28         817         742
           3 |       24         561         636
-------------+---------------------------------
    combined |       52        1378        1378

unadjusted variance      2968.00
adjustment for ties       -47.00
                       ----------
adjusted variance        2921.00

Ho: c2sflex(subpop==2) = c2sflex(subpop==3)
             z =   1.388
    Prob > |z| =   0.1652

22 . ranksum c2sflex2008, by(subpop)

Two-sample Wilcoxon rank-sum (Mann-Whitney) test

      subpop |      obs    rank sum    expected
-------------+---------------------------------
           2 |       28       787.5         742
           3 |       24       590.5         636
-------------+---------------------------------
    combined |       52        1378        1378

unadjusted variance      2968.00
adjustment for ties       -38.39
                       ----------
adjusted variance        2929.61

Ho: c2sfle~8(subpop==2) = c2sfle~8(subpop==3)
             z =   0.841
    Prob > |z| =   0.4006
```

```
3Cs Today&2008   Tuesday February 15 23:17:39 2011   Page 8

23 . ranksum c2postpone, by(subpop)

Two-sample Wilcoxon rank-sum (Mann-Whitney) test

      subpop |      obs    rank sum    expected
-------------+---------------------------------
           2 |       28       683.5         742
           3 |       24       694.5         636
-------------+---------------------------------
    combined |       52        1378        1378

unadjusted variance     2968.00
adjustment for ties     -165.97
                       ----------
adjusted variance       2802.03

Ho: c2post~e(subpop==2) = c2post~e(subpop==3)
             z =  -1.105
    Prob > |z| =   0.2691

24 . ranksum c2postpone2008, by(subpop)

Two-sample Wilcoxon rank-sum (Mann-Whitney) test

      subpop |      obs    rank sum    expected
-------------+---------------------------------
           2 |       28         710         742
           3 |       24         668         636
-------------+---------------------------------
    combined |       52        1378        1378

unadjusted variance     2968.00
adjustment for ties     -118.97
                       ----------
adjusted variance       2849.03

Ho: c2p~2008(subpop==2) = c2p~2008(subpop==3)
             z =  -0.600
    Prob > |z| =   0.5488

25 . ranksum c2abandon, by(subpop)

Two-sample Wilcoxon rank-sum (Mann-Whitney) test

      subpop |      obs    rank sum    expected
-------------+---------------------------------
           2 |       28       711.5         728
           3 |       23       614.5         598
-------------+---------------------------------
    combined |       51        1326        1326

unadjusted variance     2790.67
adjustment for ties     -153.93
                       ----------
adjusted variance       2636.74

Ho: c2aban~n(subpop==2) = c2aban~n(subpop==3)
             z =  -0.321
    Prob > |z| =   0.7480
```

```
3Cs Today&2008   Tuesday February 15 23:17:39 2011   Page 9

26 . ranksum c2abandon2008, by(subpop)

   Two-sample Wilcoxon rank-sum (Mann-Whitney) test

         subpop |      obs    rank sum    expected
   -------------+---------------------------------
              2 |       28       683.5         728
              3 |       23       642.5         598
   -------------+---------------------------------
       combined |       51        1326        1326

   unadjusted variance     2790.67
   adjustment for ties     -174.89
                          ----------
   adjusted variance       2615.78

   Ho: c2a~2008(subpop==2) = c2a~2008(subpop==3)
                z =  -0.870
       Prob > |z| =   0.3843

27 . ranksum c2rescale, by(subpop)

   Two-sample Wilcoxon rank-sum (Mann-Whitney) test

         subpop |      obs    rank sum    expected
   -------------+---------------------------------
              2 |       28         738         742
              3 |       24         640         636
   -------------+---------------------------------
       combined |       52        1378        1378

   unadjusted variance     2968.00
   adjustment for ties     -256.94
                          ----------
   adjusted variance       2711.06

   Ho: c2resc~e(subpop==2) = c2resc~e(subpop==3)
                z =  -0.077
       Prob > |z| =   0.9388

28 . ranksum c2rescale2008, by(subpop)

   Two-sample Wilcoxon rank-sum (Mann-Whitney) test

         subpop |      obs    rank sum    expected
   -------------+---------------------------------
              2 |       28       736.5         742
              3 |       24       641.5         636
   -------------+---------------------------------
       combined |       52        1378        1378

   unadjusted variance     2968.00
   adjustment for ties     -165.97
                          ----------
   adjusted variance       2802.03

   Ho: c2r~2008(subpop==2) = c2r~2008(subpop==3)
                z =  -0.104
       Prob > |z| =   0.9172
```

```
3Cs Today&2008   Tuesday February 15 23:17:39 2011   Page 10

29 . ranksum c2growth, by(subpop)

   Two-sample Wilcoxon rank-sum (Mann-Whitney) test

         subpop |      obs    rank sum    expected
   -------------+---------------------------------
              2 |       28         788         742
              3 |       24         590         636
   -------------+---------------------------------
       combined |       52        1378        1378

   unadjusted variance     2968.00
   adjustment for ties     -219.82
                          ----------
   adjusted variance       2748.18

   Ho: c2growth(subpop==2) = c2growth(subpop==3)
                z =   0.877
       Prob > |z| =   0.3802

30 . ranksum c2growth2008, by(subpop)

   Two-sample Wilcoxon rank-sum (Mann-Whitney) test

         subpop |      obs    rank sum    expected
   -------------+---------------------------------
              2 |       28         790         742
              3 |       24         588         636
   -------------+---------------------------------
       combined |       52        1378        1378

   unadjusted variance     2968.00
   adjustment for ties     -153.43
                          ----------
   adjusted variance       2814.57

   Ho: c2g~2008(subpop==2) = c2g~2008(subpop==3)
                z =   0.905
       Prob > |z| =   0.3656

31 . ranksum c2flex, by(subpop)

   Two-sample Wilcoxon rank-sum (Mann-Whitney) test

         subpop |      obs    rank sum    expected
   -------------+---------------------------------
              2 |       28       718.5         742
              3 |       24       659.5         636
   -------------+---------------------------------
       combined |       52        1378        1378

   unadjusted variance     2968.00
   adjustment for ties      -54.48
                          ----------
   adjusted variance       2913.52

   Ho: c2flex(subpop==2) = c2flex(subpop==3)
                z =  -0.435
       Prob > |z| =   0.6633
```

```
3Cs Today&2008   Tuesday February 15 23:17:39 2011   Page 11

32 . ranksum c2flex2008, by(subpop)

Two-sample Wilcoxon rank-sum (Mann-Whitney) test

      subpop |      obs    rank sum    expected
-------------+---------------------------------
           2 |       28       728.5         742
           3 |       24       649.5         636
-------------+---------------------------------
    combined |       52        1378        1378

unadjusted variance     2968.00
adjustment for ties      -53.72
                     ----------
adjusted variance       2914.28

Ho: c2f~2008(subpop==2) = c2f~2008(subpop==3)
             z =  -0.250
    Prob > |z| =   0.8025

33 . ranksum c3postpone, by(subpop)

Two-sample Wilcoxon rank-sum (Mann-Whitney) test

      subpop |      obs    rank sum    expected
-------------+---------------------------------
           2 |       29       840.5         783
           3 |       24       590.5         648
-------------+---------------------------------
    combined |       53        1431        1431

unadjusted variance     3132.00
adjustment for ties     -188.65
                     ----------
adjusted variance       2943.35

Ho: c3post~e(subpop==2) = c3post~e(subpop==3)
             z =   1.060
    Prob > |z| =   0.2892

34 . ranksum c3postpone2008, by(subpop)

Two-sample Wilcoxon rank-sum (Mann-Whitney) test

      subpop |      obs    rank sum    expected
-------------+---------------------------------
           2 |       29       869.5         783
           3 |       24       561.5         648
-------------+---------------------------------
    combined |       53        1431        1431

unadjusted variance     3132.00
adjustment for ties     -151.02
                     ----------
adjusted variance       2980.98

Ho: c3p~2008(subpop==2) = c3p~2008(subpop==3)
             z =   1.584
    Prob > |z| =   0.1131
```

```
3Cs Today&2008   Tuesday February 15 23:17:39 2011   Page 12

35 . ranksum c3abandon, by(subpop)

Two-sample Wilcoxon rank-sum (Mann-Whitney) test

      subpop |      obs    rank sum    expected
-------------+---------------------------------
           2 |       29         799         783
           3 |       24         632         648
-------------+---------------------------------
    combined |       53        1431        1431

unadjusted variance     3132.00
adjustment for ties     -188.65
                      ----------
adjusted variance       2943.35

Ho: c3aban~n(subpop==2) = c3aban~n(subpop==3)
             z =   0.295
    Prob > |z| =   0.7681

36 . ranksum c3abandon2008, by(subpop)

Two-sample Wilcoxon rank-sum (Mann-Whitney) test

      subpop |      obs    rank sum    expected
-------------+---------------------------------
           2 |       29         818         783
           3 |       24         613         648
-------------+---------------------------------
    combined |       53        1431        1431

unadjusted variance     3132.00
adjustment for ties     -189.66
                      ----------
adjusted variance       2942.34

Ho: c3a~2008(subpop==2) = c3a~2008(subpop==3)
             z =   0.645
    Prob > |z| =   0.5188

37 . ranksum c3rescale, by(subpop)

Two-sample Wilcoxon rank-sum (Mann-Whitney) test

      subpop |      obs    rank sum    expected
-------------+---------------------------------
           2 |       29         931         783
           3 |       24         500         648
-------------+---------------------------------
    combined |       53        1431        1431

unadjusted variance     3132.00
adjustment for ties     -255.70
                      ----------
adjusted variance       2876.30

Ho: c3resc~e(subpop==2) = c3resc~e(subpop==3)
             z =   2.760
    Prob > |z| =   0.0058
```

```
3Cs Today&2008   Tuesday February 15 23:17:40 2011   Page 13

38 . ranksum c3rescale2008, by(subpop)

Two-sample Wilcoxon rank-sum (Mann-Whitney) test

      subpop |      obs    rank sum    expected
-------------+---------------------------------
           2 |       29       953.5         783
           3 |       24       477.5         648
-------------+---------------------------------
    combined |       53        1431        1431

unadjusted variance     3132.00
adjustment for ties     -221.10
                      ----------
adjusted variance       2910.90

Ho: c3r~2008(subpop==2) = c3r~2008(subpop==3)
             z =   3.160
    Prob > |z| =   0.0016

39 . ranksum c3growth, by(subpop)

Two-sample Wilcoxon rank-sum (Mann-Whitney) test

      subpop |      obs    rank sum    expected
-------------+---------------------------------
           2 |       29         959         783
           3 |       24         472         648
-------------+---------------------------------
    combined |       53        1431        1431

unadjusted variance     3132.00
adjustment for ties     -618.60
                      ----------
adjusted variance       2513.40

Ho: c3growth(subpop==2) = c3growth(subpop==3)
             z =   3.511
    Prob > |z| =   0.0004

40 . ranksum c3growth2008, by(subpop)

Two-sample Wilcoxon rank-sum (Mann-Whitney) test

      subpop |      obs    rank sum    expected
-------------+---------------------------------
           2 |       29       937.5         783
           3 |       24       493.5         648
-------------+---------------------------------
    combined |       53        1431        1431

unadjusted variance     3132.00
adjustment for ties     -298.00
                      ----------
adjusted variance       2834.00

Ho: c3g~2008(subpop==2) = c3g~2008(subpop==3)
             z =   2.902
    Prob > |z| =   0.0037
```

```
   3Cs Today&2008  Tuesday February 15 23:17:40 2011   Page 14

41 . ranksum c3flex, by(subpop)

   Two-sample Wilcoxon rank-sum (Mann-Whitney) test

         subpop |      obs    rank sum    expected
   -------------+---------------------------------
              2 |       29       891.5         783
              3 |       24       539.5         648
   -------------+---------------------------------
       combined |       53        1431        1431

   unadjusted variance     3132.00
   adjustment for ties      -29.17
                          ----------
   adjusted variance       3102.83

   Ho: c3flex(subpop==2) = c3flex(subpop==3)
                z =   1.948
       Prob > |z| =   0.0514

42 . ranksum c3flex2008, by(subpop)

   Two-sample Wilcoxon rank-sum (Mann-Whitney) test

         subpop |      obs    rank sum    expected
   -------------+---------------------------------
              2 |       29         905         783
              3 |       24         526         648
   -------------+---------------------------------
       combined |       53        1431        1431

   unadjusted variance     3132.00
   adjustment for ties      -38.51
                          ----------
   adjusted variance       3093.49

   Ho: c3f~2008(subpop==2) = c3f~2008(subpop==3)
                z =   2.193
       Prob > |z| =   0.0283

43 .
   end of do-file

44 .
```

Index

C.M. Sipp and E.G. Carayannis, *Real Options and Strategic Technology Venturing: A New Paradigm in Decision Making*, SpringerBriefs in Business 31,
DOI 10.1007/978-1-4614-5814-2,

GPSR Compliance

The European Union's (EU) General Product Safety Regulation (GPSR) is a set of rules that requires consumer products to be safe and our obligations to ensure this.

If you have any concerns about our products, you can contact us on ProductSafety@springernature.com

In case Publisher is established outside the EU, the EU authorized representative is:

Springer Nature Customer Service Center GmbH
Europaplatz 3
69115 Heidelberg, Germany

Batch number: 10844675

Printed by Printforce, the Netherlands